LOOKING AT MAMMALS
A BEGINNER'S GUIDE

LOOKING AT MAMMALS

A BEGINNER'S GUIDE

by
DERRICK KNOWLTON

LUTTERWORTH PRESS
GUILDFORD AND LONDON

First published 1979

ISBN 0 7188 2382 6

COPYRIGHT © 1979 DERRICK KNOWLTON

All rights Reserved. No part of this publication
may be reproduced, stored in a retrieval system,
or transmitted, in any form or by any means,
electronic, mechanical, photocopying, recording
or otherwise, without the prior permission of
Lutterworth Press, Farnham Road, Guildford, Surrey.

*Printed in Great Britain by
Butler & Tanner Ltd.
Frome and London*

CONTENTS

Chapter		Page
1.	INTRODUCTION	9
2.	ORIGIN OF BRITISH MAMMALS	14
3.	CLASSIFICATION AND IDENTIFICATION OF THE BRITISH SPECIES	28
4.	DISTRIBUTION AND HABITATS	46
5.	FOOD	63
6.	BEHAVIOUR	76
7.	HOW TO OBSERVE	92
8.	EQUIPMENT	109
9.	MAMMALS AND MAN	115
	INDEX	127

LIST OF ILLUSTRATIONS

PHOTOGRAPHS

		Page
1.	Red-necked wallabies	12
2.	A bank vole feeding on a hazel twig	19
3.	A wild boar	22
4.	A coypu in the Norfolk Wildlife Park	25
5.	Soay Sheep in Weyhill Wildlife Park	26
6.	The pygmy or lesser shrew	33
7.	Wood mouse or long-tailed field mouse	39
8.	Common seal swimming	43
9.	Wild goats on the cliffs of Islay	44
10.	A group of fallow deer	49
11.	The little muntjac or barking deer	51
12.	Fresh mole heaves in a meadow	53
13.	Wild cat in a tree	57
14.	Water vole or rat feeding on waterside plant	59
15.	The European otter	61
16.	Red deer stag	65
17.	Hibernating hedgehog	69
18.	A polecat in Weyhill Wildlife Park	70
19.	Red fox in attentive posture	73
20.	A badger's sett in winter	74
21.	A mole emerging at the surface	77
22.	A squirrel's drey in the fork of a large tree	79
23.	Sika stags grazing in the Southern Uplands	83
24.	Tree showing the characteristic fraying by roe deer	86
25.	Common seals basking on a Hebridean skerry	89

LIST OF ILLUSTRATIONS

26.	Roe deer slots in clayey soil	97
27.	Rabbit footprints in snow	98
28.	Fallow deer droppings	101
29.	Pine cones stripped by grey squirrels	103
30.	Pine marten scrambling on rocks	106
31.	Trap for catching rats alive	110
32.	The brown rat is an opportunist feeder	113
33.	Badger gate	119
34.	An old otter trap in the Shetlands	123

DRAWINGS

Fig. 1. An Outline map of South-east England showing the 'land bridge' to the mainland	15
Fig. 2. Footprints of a number of well-known mammals and rabbit trail	94, 95

Acknowledgements

The publishers wish to thank the following for supplying photographs.

The author: pages 12, 22, 26, 43, 44, 51, 53, 70, 74, 83, 86, 89, 101, 110, 113, 119, 123.
Heather Angel, M.Sc., F.I.I.P., F.R.P.S.: pages 19, 33, 39, 49, 59, 65, 69, 77, 79, 97, 103.
Alan Beaumont, D.Phil.: pages 25, 61, 73, 98.
L. MacNally: pages 57, 106.

CHAPTER 1

INTRODUCTION

More and more people each year are becoming interested in the wild mammals of the British Isles. I do not suppose that any one can really be certain why this is so but we can make some guesses. First of all, it is probably a spin-off from the popularity of general natural history. Then without doubt it has been encouraged by books which have described the life history of individual animals and particularly by television programmes which have taken viewers into a world seldom seen by human eyes. The increase in the spread and numbers of deer must also have brought them to the notice of the many people who enjoy rambling across the countryside. This increase has been large enough to cause the highway authorities to erect specific signs warning motorists of the possibility of deer crossing the road. These signs are now a familiar sight where woodland abuts on to highways; whether motorists heed the warning is quite another matter. A small minority of people are mammal-watchers for another reason. They are knowledgeable enough to realize that the observation of mammals is going to tax all their skills and ingenuity and they eagerly accept it as a challenge.

This book is intended for beginners of all these kinds who have decided to take up 'looking at mammals' as a hobby. Whilst it will include reference to some theoretical aspects the main emphasis will be practical, amongst other things describing means of identification in the field; the parts of the country where particular animals are to be found and the type of terrain which they prefer; feeding and breeding behaviour and, most important of all, how to observe and the specialized equipment available for use.

A number of mammals are not only well-distributed over the country in their own habitats but are also abundant. Some, such as the deer, are quite large. At first consideration, therefore, it is astonishing that mammals are so seldom seen and that an encounter with one on a casual walk is regarded as an exciting event. The fact remains that mammals are not easily observed and it would be foolish to pretend otherwise.

A personal experience will demonstrate this point. I suspected that probably there were a few small rodents in the bank at the end of my vegetable garden but I had no real evidence of this until my sowing of peas failed to appear above the ground. The obvious way both of finding out and safeguarding future sowings was to set traps. To my astonishment two traps produced nearly fifty long-tailed field mice in less than six months. This trapping revealed another interesting fact. So far from the rodent population being exterminated they were barely being kept in check. What was happening was that the temporarily unoccupied ground was continually being taken over by other mice which moved in to occupy the vacant habitat.

The significance of this aspect of animal behaviour has only begun to be appreciated in comparatively recent times and it has far-reaching implications. It is now realized, for example, that it is more satisfactory if foresters do not attempt to wipe out a colony of roe occupying a particular wood. The damage done to trees is best tolerated because the shooting of bucks will only bring in others which may well cause greater damage. I do not mean that the deer population should not be controlled at all. In the absence of natural predators it must be kept in check for humane reasons, otherwise many would die of starvation. What I mean is that deer control has to be undertaken with considerable expertise in order to preserve a delicate balance, otherwise the last state may well be worse than the first. It is now considered on similar grounds that the shooting of grey squirrels is a pointless exercise.

There are several reasons why mammals are not easy to observe. One is that a number of species are nocturnal in habit and therefore can only be seen for a brief period at dusk and dawn. Another is that they have been subject to much persecution by man and not unnaturally they have reacted by treating humans with great suspicion and by becoming adept at concealment. It is probable that some species have been driven to become nocturnal by reason of this persecution. For example, badgers which have setts in remote places tend to emerge distinctly earlier in the evening than those which live close to roads or dwellings. Then, too, in contrast with birds and butterflies which make themselves conspicuous in flight, all mammals in Britain except bats are confined to the ground. Numerically, both in total population and

INTRODUCTION

in numbers of species, birds and insects are a more prominent part of the British fauna.

All of this means that 'looking at mammals' is a challenge which produces all the excitement of the chase. As with so many hobbies the underlying attitude of mind is all-important. In mammal-watching we have to develop our hunting instincts and pit our brains against the innate cunning possessed by many mammals. Since this instinct is deeply ingrained in us it is highly satisfying to exercise it even if at the end of the day we have failed in our objective of locating our quarry. After all, there is tremendous enjoyment in the search itself and unlike primitive man, from whom our instincts derive, we are not hunting for food to keep us alive. I do not want you to imagine, however, that the difficulties are insuperable. I went out recently on a bird-watching trip with hardly a thought of mammals in my mind. Without conscious effort on my part I was able to watch a vixen hunting and obtain good views of no less than three species of deer.

Various species of mammals have differing habitat requirements and these will be described later in the book. You will do well to supplement this background information with many field observations of your own. Gradually a store of knowledge will be built up which will pass down into the subconscious. Then when you go out into the countryside and look at the surroundings there will emerge what you may think of as a sixth sense which will tell you that this is an ideal haunt for such and such an animal. It will be very likely in these circumstances that the mammal in question will indeed be living there because animal populations tend to spread out and occupy suitable territory. There may be certain limitations to this principle in practice, especially where mammals are concerned. For example, a small population of a certain species may be prevented from expanding into similar habitat some distance away by a large stretch of quite unsuitable terrain lying between. But the general principle still stands; it has been well said that 'Nature abhors a vacuum'.

The most striking personal experience which I have had of the operation of this so-called sixth sense as applied to mammals happened one summer evening. I was walking with a friend when we came to a disused and overgrown chalk pit. There were elders and clumps of nettles growing on the slopes and floor of the pit. 'Just the place for badgers,' I remarked to my friend. Immediately there was a crashing noise in the

1. Red-necked wallabies in Weyhill Wildlife Park

undergrowth and dashing forward we were just in time to see a badger scurrying away.

'Looking at mammals' is a solitary occupation because silence and a stealthy approach are essential. At most, one congenial companion who knows how to keep quiet can accompany you. It is often pleasant to know that such a friend is nearby when standing at night-time in the darkness of the woods at the entrance to a badger sett. Despite the solitary nature of much mammal watching this by no means requires the observer to be a 'loner' cut off from his fellow naturalists. It is advantageous to join the local natural history society. This will not only prevent isolationism but will keep you in touch with other branches of nature study. These local societies will often arrange for small groups to go badger watching and on deer excursions.

The Fauna Preservation Society is concerned with the conservation of mammals throughout the world. It publishes a journal entitled *Oryx* which carries articles on British mammals from time to time. The Mammal Society promotes the study of British mammals including the organizing of co-operative surveys and publishes a journal *Mammal Review*. *Countryside* is the magazine of the British Naturalists' Association and this occasionally carries articles on mammals; the contributions in this magazine are often well-suited to the needs of the beginner. There is another very specialized national body, the British Deer Society. This serves the interests of all who are concerned with deer—landowners, stalkers, foresters, mammalogists and just plain deer-lovers. The objects of this Society are to exchange information on all aspects of deer behaviour and to promote the conservation and humane control of deer. A journal entitled *Deer* is published twice yearly.

CHAPTER 2

ORIGIN OF BRITISH MAMMALS

IF we give any thought at all to this question of origin we probably regard British mammals as always being there, reproducing themselves for generation after generation from time immemorial. To think like that, however, is to fail to take into account changes in the climate and in the physical geography which have occurred in geologically recent times. The kinds of animals which will live in any country depends primarily upon the climate. You cannot expect to find arctic mammals in the tropics or vice-versa. The British Isles have not always had temperate conditions. In the era immediately before the present one many tropical mammals such as elephants, lions, rhinoceroses and hippopotamuses roamed the land.

Then the current era began with the devastating physical event which is known as the Ice Age. This lasted for nearly a million years although it was not continuously cold throughout that time; there were probably three intervening periods when the weather was milder. For a good deal of the time, however, much of the land was covered by glaciers; the arctic with its blizzards and its ice had come to Britain. Probably many of the warmth-loving animals fled south and escaped but some stayed on and succumbed. Their remains have been found in glacial deposits and in caves where they had sheltered from the cold. Across the frozen North Sea came the invaders from Scandinavia and the Arctic Circle, those mammals which were accustomed to conditions of extreme cold. These included large species such as the mammoth, musk-ox, woolly rhinoceros and reindeer. There were also smaller creatures like the mountain hare and arctic fox.

When the climate became warmer in the inter-glacial periods these would have retreated north and been replaced by the mammals of the southern plains. There must have been times when many representatives of both types were present, giving an incredibly rich variety that would have made 'looking at mammals' a delight if there had been any naturalists around at the time.

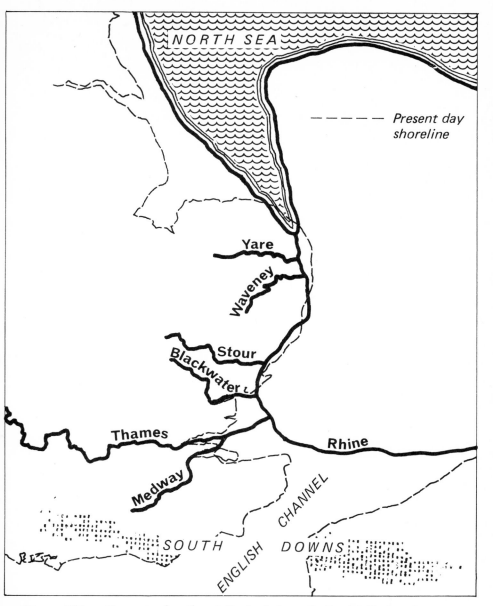

Fig. 1. This outline map of south-east England shows the 'land bridge' to the mainland before it was severed in the Ice Age

Mammals of course do not possess the mobility of birds and this fascinating alternation of cold and warm types was only possible because Britain was still well and truly joined to the Continent so enabling land migrations to take place. The final separation of Britain from the mainland of Europe cannot be dated with any certainty but it was probably around 5000 BC when the last land links were engulfed by the waves and the waters of the Atlantic rushed through the crumbling chalk to unite with the North Sea. The fact that this land bridge was still in existence at the end of the Ice Age was of vital significance to the establishment of the present mammal fauna. As the climate slowly improved so the vegetation changed to marshy tundra and then to dry steppes. Throughout this time animals were able to move northward and regain a foothold in Britain, gradually spreading out over the country wherever suitable terrain occurred. This was the principal method by which the fauna of the present day came into existence.

In the next chapter mammals will be placed in their scientific classification but they can be grouped in another way in four categories: those that re-occupied the country at the end of the Ice Age and which have remained identical to those of their species living on the Continent; those which have evolved into sub-species through island isolation; those which have become extinct and lastly those which have been introduced by man.

Quite a number of today's mammals come in the first category. They are fairly well-spaced over the country and as yet show no evidence of departing from the type which occurs on the Continent. Examples of this group include the hedgehog, mole, rabbit, weasel, fox, badger and otter. In terms of evolutionary time, however, only a very short period has elapsed and it is possible that some of these may yet evolve into British sub-species. After they arrived the climate changed from a more extreme kind of cold winters and hot summers to milder, moister conditions. This process was considerably helped by the formation of what we now call the English Channel. The majority, perhaps all, of the mammals appear to have adapted themselves quite well to the changing weather. Those that became extinct may well have done so for reasons other than climatic ones.

It is not at all easy to determine how many animals come in the second category because there is not entire agreement on which are genuine sub-species. Scientists who study the structure of animals are called taxonomists and they can themselves be separated into two classes, if not exactly two sub-species! On the one hand there are those who are prepared to name a sub-species on the evidence of slight structural divergence and these are colloquially known as 'splitters'. Those on the other hand who are not eager to sub-divide a species until there are very strong grounds for doing so are 'lumpers'. Despite the fact that the ordinary field naturalist is not really in his element in this specialized subject, it is a very fascinating one because we are looking at evolution in action. We will look first at those mammals which have evolved into just one British sub-species.

The water shrew found in Britain differs only slightly from that on the mainland of Europe. It is not quite as sharply defined black above and white below because the white underside has faint brownish patches, hence the sub-specific name allotted to it is *bicolor*. We can imagine this aquatic animal arriving in Britain along the extended course of the river Rhine which in those days, instead of ending at Rotterdam, meandered across East Anglia and entered the North Sea between Flamborough Head and the Dogger Bank which was then dry land. This tiny mammal could then have diverted into the west bank tributaries such as the Thames and spread westwards across the country. It must be said, however, that the water shrew is not confined to water and it is conceivably possible that it made its way overland into Britain.

The brown hares in Europe are light brown in colour whilst those in this country are darker brown. On this rather slender basis the latter has been differentiated as the western form of the European hare. In the same Order is another mammal where colour is the main factor in separating the British sub-species. This is the red squirrel which is distinguished from the Continental races mainly by its tail which is cream or straw-coloured during the summer and is consequently sometimes known as the light-tailed squirrel; it is also smaller than the other sub-species.

The yellow-necked mouse is a mammal which a non-specialist would probably quite happily accept as a long-tailed field mouse seeing that the latter often has traces of yellow on the throat and that the yellow spot on the former can be quite small. There are, however, slight differences in the skull which together with other small features have caused the taxonomists to determine it as a separate species. Nevertheless the two species sometimes interbreed and we can regard the yellow-necked mouse as an example of evolution in progress. The process continues with the British form, with its clear-cut yellow spot often elongated on each side, being accepted as a separate sub-species.

The last two mammals in this group are concentrated in Scotland and therefore have been given sub-specific names which indicate their origin. The Scottish wild cat is very variable in colour and size and is merely a geographical race of the so-called European wild cat which actually occurs also in Asia and Africa. Unfortunately, it is certain that many of the Scottish animals show traces of inter-breeding with domestic cats. The true wild cat has only one breeding season early in the year but in Scotland there are two litters born each year and sometimes even three. Whilst the domestic cat is most punctilious in scraping a hole and covering its droppings the wild cat often makes no attempt to cover them and in this respect at least the Scottish cats show their wild nature. The native red deer of the British Isles have been given the sub-species name *scoticus* because today these are mainly found in Scotland and the north of England. It is probably impossible now to say with any conviction which if any of the other red deer are truly native although those on Exmoor and the Quantocks may well be. But many of the red deer in England derived originally from animals which escaped from parks. The British sub-species is rather darker in colour and generally smaller than those on the Continent.

There are three mammals which have separated into two sub-species in Britain. In addition to the common shrew of the mainland there is on Islay in the Inner Hebrides an island race in which the lighter colour of the underside extends higher up the flanks than in the mainland animals. The blue mountain hare occurs in two separate forms, one in the Scottish Highlands and the other in Ireland. The latter is reddish brown and slightly larger, being nearer the Continental races in size. The third mammal is the water vole which in the Highland regions of Scotland exists in a slightly smaller race; more noticeable to the observer in the field will be the darker colour. Black water voles are very common in East Anglia and one would naturally think that these might well belong to the Scottish sub-species but it is considered that these are merely black forms of the principal British sub-species.

The voles, field mouse and stoat in Britain have evolved into several sub-species. Isolation is the key factor involved. This isolation may arise

2. *A bank vole feeding on a hazel twig*

from separation on islands, or other physical or climatic barriers. Probably all these factors have played their part in the development of sub-species in the small mammal population but most of the races of the short-tailed vole have originated on the Hebridean islands. At the end of the Ice Age short-tailed voles moved north over the country and this original population now occupies high ground in Scotland. Later another wave entered the country and took over the lowlands and this has been designated as the English sub-species although it occurs also in the lowlands of Scotland. Most of the islands of the Inner and Outer Hebrides have the Hebridean sub-species but on four, Islay, Gigha, Muck and Eigg the voles have developed sufficiently to be accorded sub-species status. It is interesting to note how close these two pairs of islands are to each other; Islay and Gigha about 15 km (9 miles) apart and Muck and Eigg much less, separated only by the 5 km (3 miles) wide Sound of Eigg yet they have all evolved into individual races whilst islands as far apart as North Uist and Mull share the same Hebridean vole. The various races are distinguished by slight differences in colour, density of fur and tooth structure.

I can confirm from my own experience that the darker colour of the vole in the Orkneys can be detected by the observer in the field and it has been made a full species in its own right. As in the Hebrides a number of the Orcadian islands have developed their own sub-species of this vole.

Four races of the bank vole occur in Britain. Apart from the sub-species on the mainland there are separate races on the Hebridean islands of Raasay and Mull; the fourth is the most surprising, however, for it is found much further south on the small Welsh island of Skomer where it is reputed to be easily approached. It is considered that these island races originated not from the mainland sub-species but from an older form which arrived in Britain during one of the earlier inter-glacials in the Ice Age. The evidence for the presence of this race in Britain long ago exists in fossil remains.

When we come to the field mouse we find that no less than seventeen different sub-species have been described; these are mainly on the northern islands ranging from Berneray in the southern part of the Outer Hebrides to the Shetland island of Foula in the far north. The differences are so minute that they would not concern anyone other than specialists and even these are not in agreement on whether sub-specific status for so many is justified.

The last species to be mentioned in this category is the stoat which has three races in the British Isles, on the mainland, on Islay and in Ireland. It is surprising that if this mammal is differentiating into subspecies it should do so only on one Scottish island; the Islay form is rather smaller but from my own observations this is not discernible in the field.

The whole question of divergence within species is full of speculative problems. Why, for example, do some mammals remain quite constant in type whilst others readily evolve as soon as a small population becomes isolated? Why should a separate race develop on one island but not on a neighbouring one? We do not know the answers to these and kindred questions. What we do know and what is of most interest to the field naturalist is that Nature is never static but always dynamically evolving to adapt to the specific conditions of the environment.

Not all of the animal species that re-entered Britain at the end of the Ice Age were able to survive; some became extinct either through loss of habitat or extermination by man. In the mild and wet climate which prevailed about 6000 years ago, vast deciduous forests covered the country. In them and on the plains wandered herds of European bison and wild cattle; the latter, known as aurochs, were huge red brown beasts standing 1·8 m (5 ft. 9 in.) tall and possessing black-tipped horns. Both kinds provided good meat and clothing for Stone Age men and so they were eagerly hunted. We know that they were attacked by New Stone Age men because a stone axe has been found fastened in the smashed skull of an aurochs. This hunting may have been overzealous for by the time of the Celts in the Iron Age both species had become extinct in Britain.

The brown bear was also appreciated for the meat it provided and this mammal is believed to have lingered on until Saxon times. We can well imagine that the bear was nothing like so easy prey as the wild cattle for it is a cunning beast and better able to escape its pursuers. Early man would also have treated this powerful animal with respect although it was not likely to have been dangerous until provoked or cornered.

One more animal which supplied the Celts and Saxons with meat was the wild boar whose natural habitat was the large oak woodlands where they fed greedily on acorns. The Norman kings hunted them

3. A wild boar in Weyhill Wildlife Park

for sport but by this time their numbers must have been decreasing. William the Conqueror made a law decreeing blindness as the penalty for killing a wild boar and this brutal punishment probably indicates that these animals were becoming quite rare. They are known to have been still in existence in several forests during the 14th century but it is probable that by the end of this century they were virtually extinct in England. In Scotland they survived until the 17th century and in 1830 a number were released in the Forest of Mar on Deeside but they did not survive. On forest land near the town of New Galloway a cairn stands on the spot where the last wild boar in Scotland is supposed to have been killed. There are those who would like to see them once again roaming through the trees but whether that will ever come to pass is another matter.

Elk and reindeer were widespread in the early days after the Ice Age and were very much at home on the spacious tundras. There is evidence of mass migrations of the latter animal in southern England but when the deciduous forests spread over the land the habitat and climate became unsuitable for the reindeer and the climate for the elk, so both species moved north to the colder climate and more open country of Scotland. The elk probably disappeared first but it is known that the reindeer was hunted on the bleak plateau of Caithness in the 12th century. Then it, too, vanished from the scene until the 18th and 19th centuries when two unsuccessful attempts were made to bring reindeer back. An experimental herd was reintroduced in the Cairngorms in 1952 and this has been more successful.

The early herds of these deer were preyed upon by packs of wolves which thrived in the forests. The reduction of the woodlands and increasing persecution drove them to the wilder parts of Britain and there they held out in dwindling numbers until the 18th century. A stone near the roadside at Lothbeg in Glen Loth in eastern Sutherland records the death of the county's last wolf in 1700. It is believed that Scotland's last wolf met its end at the hands of a hunter in Morayshire in 1743.

Beavers were active on the rivers in early historic times but their dam building caused flooding and with increasing human population they were exterminated in England although they are reported still to have occurred in Inverness-shire as late as the 16th century. An attempt may shortly be made to re-establish beavers in this country by bringing in a few pairs from France.

After a gap of nearly 200 years another mammal became extinct. This was the St. Kilda house mouse originally designated a full species but really a sub-species of the mainland mouse. Like so many of its kind it lived in close association with and dependence on the St. Kildans, a state to which biologists give the name 'commensal'. When the islanders were evacuated in 1930 the unfortunate mice, deprived of their main sources of food, had to forage abroad where they came into competition with field mice and within a few months they became extinct.

The fourth category covers those mammals which have been introduced by man. We will not concern ourselves here either with short-lived attempts or with those that have merely been transferred from one part of the British Isles to another, but only with those animals which have been successfully established as part of the British mammalian fauna whether deliberately or otherwise. It may come as a surprise to many people to learn that neither rats nor mice are really native to Britain. It is virtually certain that all three species came in the first place from Asia as stowaways on ships. Over the centuries they have developed from a wild existence in the open to living in or around buildings in close association with man where they successfully exploit the abundance of food. It is not known when the common house mouse arrived in this country but it has been here for many hundreds of years and was the first of the three to come. The black rat is sometimes described as the original British rat but it arrived sometime in the 13th century possibly on the ships carrying the Crusaders back from Palestine. This creature was responsible for the Black Death and other mediaeval outbreaks of plague. It seems that there was a westward movement of the brown rat from eastern Asia across land to Europe and ships finally brought the animal to Britain early in the 17th century where it has replaced the black rat as the common species.

Another method of accidental introduction is when animals escape from captivity in parks, zoos or fur farms. Rabbits have existed in this country for hundreds of years yet they are not native animals. They were brought into Britain during the Norman conquest. It seems certain that they were kept in warrens for food and for their fur but inevitably some escaped and gradually spread through the countryside. The changes in agriculture which took place from the 18th century onwards doubtless benefited the rabbit population and they became abundant in many places until the arrival of the terrible disease of myxomatosis.

Mink began escaping from fur farms in the 1930's but it was not until the mid 1950's that breeding in the wild was known to have taken place. This was along the banks of the river Teign in Devon. Other escapes from various fur farms have continued and the mink is now widespread throughout the British Isles, even occurring in the Outer Hebrides.

Another escape of this type and at about the same time is that of the coypu, a South American rodent which is bred for the fur known as nutria. The establishment of this mammal, however, followed a quite different pattern from that of mink. Both are aquatic animals but coypu appears more restricted to swamps and fens. They began breeding successfully in several counties but a concentrated campaign against them has eliminated all outlying colonies, leaving just one main area of occupation in Norfolk and eastern Suffolk. It is generally recognized that here they will have to be accepted as permanent residents.

The red-necked wallaby is now a member of the British fauna and a totally unexpected one, for who would have thought that an Australian mammal would breed in the wild in this country. This particular species, however, is quite hardy and several colonies now exist, in Sussex, the Peak District and on an island in Loch Lomond.

The grey squirrel from America was released in a number of places including Regents Park in London and Woburn in Bedfordshire mainly during the latter part of the last century and these animals quickly found an available niche in deciduous woodland. They have now spread to many parts of the country, becoming a pest to the forester. Despite a vigorous shooting campaign they still thrive and there is no question of their ever being exterminated by this means; the grey squirrel, pest or not, is here to stay.

About the end of the 19th century Lord Rothschild released some

4. A coypu in the Norfolk Wildlife Park

5. *Soay Sheep in Weyhill Wildlife Park*

specimens of the fat or edible dormouse on his estate in Tring, Hertfordshire. Its natural habitat is similar to that of the grey squirrel but in Britain it seems to live principally in buildings, at least in wintertime. Just as the coypu might have been expected to spread like the mink but did not, so the edible dormouse in theory ought to have been able to expand its territory similarly. Instead it occupies the Chiltern countryside and nothing more.

Several species of deer have been released in various places and are now firmly established. The earliest introduction is that of the fallow; the origin of this deer is in some doubt but it is popularly believed that they were brought here by the Romans although it is possible that the Normans were responsible. There have been numerous escapes from deer parks to augment those living in the wild especially during the Second World War when fences were not maintained and deer were in any case allowed to wander off and fend for themselves in order to save the cost of looking after them. The sika deer is a native of Japan and was introduced here about the turn of the century. They exist in a feral state in a number of counties but generally they are restricted to localized areas and do not roam far and wide like some other species. One of these is the little muntjac known as the barking deer which spread originally from the Woburn estate and now is found in a number of counties although the greatest density is in the Chiltern countryside and the south-east Midlands. It is a great wanderer and individuals are liable to appear almost anywhere. Two more species, the Siberian roe and the Chinese water deer, live in the wild in very limited numbers and both also stem from the Woburn estate.

Two domestic species live here in a feral state. The origin of the Soay sheep which live on the remote island group of St. Kilda is shrouded in mystery. Soay is a Norse word meaning sheep island but whether that means the sheep were brought there by the Vikings or whether they named the island because the sheep were already there, brought possibly by Bronze Age folk, we do not know. What is certain is that they belong to a very primitive breed strongly resembling the sheep of New Stone Age times. Feral herds of goats live in a number of places, mainly in mountainous areas and on precipitous cliffs where they are incredibly sure-footed. They must have originated from stock imported into Britain in very early times.

Deliberate introductions of alien mammals are not acts to be lightly undertaken. There is a grave danger that the intruder will disturb the delicate balance of Nature and become a serious pest. Releases of an alien species are really only justified if three conditions are fulfilled. One is that they were once part of the native fauna and have only become extinct in historic times. The second is that the climate is suitable; it would be pointless to try and reintroduce an animal that was on the edge of its range and in this connection we have to realize that long-term climatic changes are going on of which we may be unaware but to which the animal world is keenly sensitive. The third condition is that a sufficient quantity of suitable habitat is available where the animals will be reasonably free from disturbance or persecution. If these conditions can be met then there may be no objection to small, cautious, carefully planned and controlled introductions but anyone contemplating such a course of action needs to think more than once and certainly ought to obtain the most expert advice available.

CHAPTER 3

CLASSIFICATION AND IDENTIFICATION OF THE BRITISH SPECIES

BEGINNERS in any natural history subject are not usually very concerned about scientific classification and understandably so. Those who wish to look at mammals, for example, are keen to see living ones in the field and are not especially interested in knowing to which particular genus an animal belongs. Nevertheless, an elementary working knowledge of the families of mammals can be of practical value to the budding mammalogist. Animals within the same family may not only have structural resemblances but also share certain mannerisms as well. An example of this is the notable inquisitiveness of the members of the weasel family. To be able to recognize family likeness is bound to be of use to the field naturalist in helping identification. In actual practice this ability will usually be exercised subconsciously without the observer being aware of the fact.

Classification is man's attempt to reproduce in a set form the order and constancy that lies in Nature. Because it is necessarily arbitrary and artificial it is occasionally subject to change as new facts about a species comes to light or it is ascertained that another name had been given to the animal at an earlier date. The diagram opposite shows how the groupings are arranged. It would, of course, be far too clumsy to include all the different categories in the scientific name of each animal so in practice only two names are normally used. The first is called the generic name and this indicates the genus to which it belongs. The second is the specific name and this, together with the former, identifies the mammal. Occasionally a third name is added when the species itself is divided into sub-species.

Let us now see how this works out in practice by looking at one concrete example, the short-tailed field vole which is found on the Inner Hebridean island of Islay. This is obviously an animal with a backbone and skull so that it must belong to the sub-phylum Vertebrata. On

CLASSIFICATION AND IDENTIFICATION

Diagram of Mammal Relationships

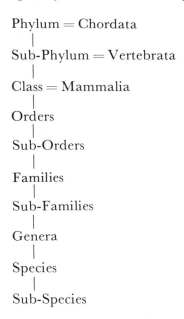

examination it is seen to have several features including, in the female, the ability to secrete milk which places it in the class Mammalia. It is fairly easy also to recognize that it bears some resemblance to rats and mice so that it is a rodent; this places it in the order Rodentia. But this vole clearly shows an even closer relationship to the other voles, in other words, it belongs to the vole family Cricetidae. This is a very large family indeed including foreign creatures like pet hamsters, and the voles occurring in Britain are all grouped in the sub-family Microtinae. Within this sub-family is the genus of the field voles, *Microtus*. In this genus is the short-tailed field vole *M. agrestis*. But we have still not reached the end of the road, for the short-tailed field vole on Islay has certain small points of difference from the mainland voles. There has to be another division into a separate sub-species so that the Islay vole becomes *M. a. fiona*.

You may wonder why scientific names are used at all seeing that all British mammals have a common name. The reason is that the scientific name is used throughout the world and naturalists of all countries can know which animal is under discussion. This is very important because many different local names are given to the same mammal. Even in Britain, for instance, in the Shetland Isles a stoat is called a weasel and vice-versa, which is confusing to say the least!

There follows now a classified list of the British mammals covered in the text together with a brief description of the field characters which are useful aids in identification. As this book is intended for beginners no reference will be made to the marine mammals other than the two breeding species of seals. Observation of the other marine mammals is infrequent and identification extremely difficult, even for an experienced mammalogist, since often the only good views obtained are when one is stranded on the shore.

As we have seen in the previous chapter there is one mammal in the order of marsupials breeding in the wild in this country. This is the red-necked wallaby *Macropus rufo-griseus*. This alien from Australia is a member of the Kangaroo family and it could not possibly be confused with any other British mammal with its upright stance, its enormous hind feet, its pouch and long, thick tail. It has a greyish brown back with a rust-red area on the neck and a white chest and belly.

The first order of native mammals is that of the insectivores which contains three families, the hedgehog, mole and the shrews. The hedgehog, *Erinaceus europaeus* is so well known that it hardly requires a description; its spine-covered back immediately identifies it. The name hog has been given because the shape of the head resembles a pig's. Country people sometimes call the hedgehog an urchin and the origin of this name goes back nearly a thousand years to the time of the Normans who used the name herichun which became corrupted to urchin. Hedgehogs of this species are well-distributed in Europe and over much of Asia. They belong to one of the oldest mammal families still in existence for they were present some 40 million years ago in very similar form to those of today.

The mole *Talpa europaea* is a burrowing animal living much of its life underground; although it does occasionally appear above the ground it is seldom seen by the public in general. Despite this the mole is well-known, perhaps because older people have seen the mole-catchers' gibbets in the countryside and because children have seen pictures of moles in their story books. Even those who are not familiar with the animal will have seen the mole-hills which are such a common sight in the fields. For those who do not know what a mole looks like, it is an animal about the size of a rat, with dark grey or black velvet fur, occasionally with a yellow tinge, a conspicuous snout, short tail and legs, the front ones being broad and shovel-like. There are a number of other members of this family in Europe, Asia and America but no more in Britain so there should be no problem in identification.

The last family in the order Insectivora is that of the shrews. When the vernacular name of an animal includes the word 'common' it does not always follow that it is widespread and plentiful. In the case of the common shrew *Sorex araneus*, however, the name aptly describes its status in Britain for it is certainly the commonest of the shrews. The name derives from a Saxon word meaning to tear which presumably refers to its aggressive manner in attacking its prey with its many sharp-pointed teeth. Local names given to it in certain parts of the country include blind mouse and red-toothed shrew. This species is not much more than 100 mm (4 in.) from head to tail and has a warm brown back with lighter-coloured belly. If you get a reasonably good view of this small animal the pointed snout will immediately identify it as some kind of shrew. These mammals can often be heard squeaking as they scuttle about amongst the vegetation and if you possess acute hearing this chattering will help to confirm the identification. The next problem is to which species does it belong. If the tail is seen to be about half the length of the body and head and in addition the sides of the body have a narrow band of pale brown contrasting with a darker brown back and a grey belly with a slight yellowish hue, then the animal is a common shrew.

The pygmy shrew *Sorex minutus* may easily be confused with the common shrew until experience is gained. The total length is about 2 cm less than the common species but the tail is proportionately longer, being at least two-thirds the length of head and body. Note that it is the proportionate length that we are talking about because the actual length of the tails of the two species is about the same. There is no intermediate colour along the sides of the body in this species, and the body colour tends to be greyer than that of the common shrew. Large numbers of shrews are found dead in the autumn and this gives an opportunity to gain experience of the different species. If the length of the skull is 16 mm ($\frac{5}{8}$ in.) or less it will be a pygmy shrew; if 18–20 mm

($\frac{3}{4}$in.), a common shrew. Although shrews are insectivorous they will enter live mammal traps. It must be remembered, however, that shrews cannot live long without food and warmth so that traps should be inspected as frequently as possible, say every two or three hours. Country people in Scotland have the delightful name, 'wee thraw', for the pygmy shrew.

It must not be thought that because a shrew is found on land that it cannot therefore be a water shrew because this mammal will at times wander well away from water. *Neomys fodiens* is the largest of the shrews in Britain and is readily recognized by its densely furred, velvety black back and silvery white underparts. The underside of the tail has a keel of stiff hairs and the feet also are fringed with hairs; these are valuable aids in swimming and diving. The head and body length varies from 70–95 mm ($2\frac{3}{4}$in. to $3\frac{3}{4}$in. long) and the tail is about three-quarters of this. In parts of England the water shrew is called the shrimp mouse because of its fondness for freshwater shrimps.

All the foregoing species of shrews have red-tipped teeth. The lesser white-toothed shrew *Crocidura suaveolens* is distinguished by the absence of the red tips. It is sometimes called the Scilly shrew because in Britain it only occurs in the Scilly Isles although it is widespread in much of Europe and Asia and occurs on two of the Channel Islands. The size of this mammal comes between that of the pygmy and common shrews and overlaps both species. The colour is grey-brown on the upper parts and a lighter grey on the underside. No other species of shrew is found on the Scillies so that makes it easy for the beginner who is looking at mammals.

So much for the insectivores. We come now to another order, that of the bats, Chiroptera. The fact must be faced that bats are among the most difficult of mammals to identify when they are on the wing. They can be identified by their calls but this requires expert knowledge and the use of special equipment. They can be seen sleeping during the day or hibernating during the winter, amongst other places on the walls and roof of caves, but they should not on any account be disturbed in these circumstances. It has been necessary to put grilles over the entrance to certain caves which harbour bats in order to protect them from human disturbance and two species, the greater horseshoe and the mouse-eared, receive full legal protection under the Conservation of Wild Creatures and Wild Plants Act, 1975.

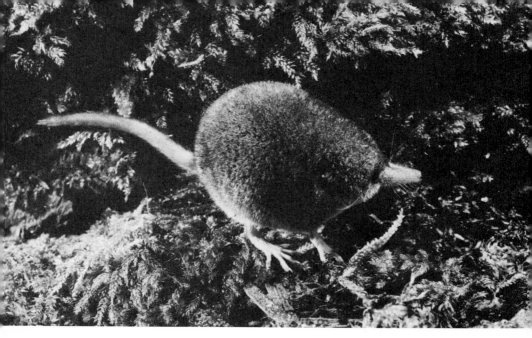

6. *The pygmy or lesser shrew*

The difficulty of making exact identifications need not by any means deter the beginner from enjoying himself if he is fortunate enough to live near a colony of bats. With the vast majority of wild creatures, accurate identification is strongly desirable and to the naturalist with a scientific bent it is an essential part of the enjoyment. Where bats are concerned a different outlook is required. We have to reconcile ourselves to the fact that we must be prepared to put a question mark after our identifications and not be ashamed to do so. That cautionary word having been said we can go on to recognize that bats vary in size and in mode of flight, and tentative identifications can be made based on these varying characteristics. The observer needs eyesight that adjusts readily to twilight conditions; not everyone has this ability but to those who have, great fun can be obtained in looking at bats.

There are 15 species of bats living in Britain. Thirteen of them belong to one family but two species, the horseshoe bats, are contained in a separate family. These last-named are distinguished by their different ear and nose structure. The name derives from the pattern of the nose which has folds of skin resembling a horseshoe. The other bats have a growth called a tragus within the ear, looking like an inner ear, but this is not present in the horseshoe bats. The greater horseshoe *Rhinolophus ferrumequinum* is one of the largest British bats and the broad, round-tipped wings have a span of approximately 350 mm (14 in.). The flight, although somewhat heavy, is graceful with alternating wing-flapping and gliding in the style of a butterfly. This bat almost always flies quite low over ground or river although exceptionally it may rise to mid-tree level in mature woodland.

The lesser horseshoe *R. hipposideros* is a distinctly smaller mammal the wing span being two-thirds of the previous species. The flight differs markedly from that of the greater horseshoe in that the wing beats are extremely rapid and sudden twists are made.

We look now at the family which contains the remainder on the British list. The whiskered bat *Myotis mystacinus* is a small bat no bigger than the little pipistrelle. The colour is variable being sometimes yellow-brown, sometimes grey on the upper parts and white underneath. A small bat seen flying in the daytime may well be this species since it frequently indulges in this habit. As well as being similar in size to the pipistrelle it also flies in a similar manner, working a regular beat around buildings or along a hedge or line of trees at a fairly fast speed. How then can it be distinguished? If you possess exceptionally acute hearing but cannot hear any sound being emitted this may be a pointer to the whiskered which is always silent in flight whereas the pipistrelle is continually calling. Provided that the animal is not flying too low the white underside may be seen. Unlike the pipistrelle which takes flying insects, the whiskered bat often hunts along fences and foliage catching spiders and insects. If a dead specimen is picked up the facial whiskers on which its name is based, will be seen.

Brandt's bat *M. brandti* is an illustration of the fact that species new to Britain are still being found and an incentive to all who look at mammals to be constantly on the lookout for anything different. This is very similar to the foregoing species from which it has small anatomical differences; they cannot be separated in the field although the colouring of adult Brandt's is reddish brown. It was only identified as a separate species on the Continent in 1970 and since then it has been discovered in a number of places in Wales, the Midlands and southern England.

CLASSIFICATION AND IDENTIFICATION

Natterer's bat *M. natteri* is medium-sized and very variable in colouring. The ears are widely spaced and have a narrow thinly-pointed tragus. The head is densely furred. The part of a bat's wing which is situated between the legs and forms a pouch is called the inter-femoral membrane. The edge of the membrane in this particular species is fringed with hairs and this is what is termed a diagnostic character, that is, it is adequate to determine the species. Of course, you would have to hold the bat in your hand in order to see this so that this feature is only of use if you should pick up a dead or injured Natterer's. More helpful characters for the field naturalist are the relatively short tail which is often held stiffly horizontally and the wings which in flight seem lighter coloured than the body. The flight is steady, not erratic. When flying slowly the tail may be depressed. Natterer's calls continuously in flight and for a bat the voice is loud.

The next two need not concern us much. Bechstein's *M. bechsteini* is a rare species of very local occurrence. The mouse-eared *M. myotis* has only recently been discovered in Britain and only two small colonies are known, both in southern England. It is the largest British bat with a wing span up to 450 mm (18 in.). It often does not emerge before darkness has fallen which is obviously a great handicap to observations.

Daubenton's *M. daubentoni* is sometimes called the water bat because it is often seen hunting over water. It is larger than the pipistrelle with which it is often confused. In this species the ears are set well apart and the inner ear is markedly convex on its hind edge. The flight is deceptively fast and usually at a low altitude. Over water it is sometimes caught on the hooks of fishing rods. Its voice is rather lower pitched than the majority of bats.

With the serotine *Eptesicus serotinus* we come to another genus with only one member in Britain. It is a large bat with a wing span equalling that of the greater horseshoe and the noctule. If you held one in your hand you would not fail to see the noticeably large teeth and you would make sure that you kept your fingers well away from them. The ears are large and broadly triangular and the inner ear is short with a rounded tip. The tail projects well beyond the membrane. On the wing the serotine is usually slow although it can fly fast on occasion and its steep dives are characteristic.

Two species of the genus Nyctalus occur in Britain. One is Leisler's *N. leisleri*. This appears to be a very local and probably uncommon mammal in England and Wales and it does not occur in Scotland. The flight is fast and fairly level and a distinguishing characteristic is reported to be a deep downward beat when the wings are brought well below the body. The other species, the noctule *N. noctula* is much more likely to be seen than Leisler's. It is one of the three large British bats and is distinctly heavy bodied. Noctules, like serotines, make steep dives but the former are generally to be separated by the fast, direct flight at a high level.

The bat most likely to be seen by a beginner is the pipistrelle *Pipistrellus pipistrellus* which is not only common but widespread in the British Isles. It is the smallest British bat and flies in an irregular manner at roof-top height or below. The barbatelle *Barbastella barbastellus* is of moderate size and local in distribution. It flies at about the same height as the pipistrelle but in a rather heavy manner. If a dead specimen is found it can be identified by the very broad short ears which virtually touch each other at the base.

Ears are a distinguishing character of the next species, the long-eared *Plecotus auritus* which has very conspicuous ears which are longer than the body itself. It is of rather small size, very common and well-distributed. The flight is seldom high and it is butterfly-like with frequent hovering. The close relative, the grey long-eared *P. austriacus* has only been identified in England in recent years and has only been found in three southern counties.

The order Lagomorpha which is now listed next to the bats is a very different one and contains only three mammals found in Britain. These are rabbits and two kinds of hare, comprising one family, the Leporidae. The rabbit *Oryctolagus cuniculus* is such a familiar sight, even after myxomatosis, that a detailed description is not necessary. The field observer has only one problem and that is to separate them from hares. This, however, is not difficult because for one thing rabbits are smaller with noticeably shorter ears and for another, the proportionately shorter hind legs give a different style of running from the hares. Cony is an old name for rabbit which has now largely passed into disuse; rabbit itself originated from a French name given to the young.

Hares belong to the genus *Lepus* the brown hare being *L. capensis*. Unlike the rabbit the brown hare is not a sociable creature although several may be seen together in the spring. The conspicuous black tips to the ears are a distinguishing feature and as already indicated, these ears

CLASSIFICATION AND IDENTIFICATION

are longer than in the rabbit. The exceptionally long hind legs give it a very characteristic appearance when running. The male is known as the jack and is rather smaller than the jill.

The blue mountain hare *L. timidus* is a separate species occurring in Scotland and parts of northern England. Ireland is occupied by a sub-species *L. t. hibernicus* which does not have a winter white coat. Both the adjectives in the common name are rather misleading. The mountain hare is found on quite low ground in some parts and the colour is generally greyish-brown in summer, certainly not blue. In winter it moults to mainly white, although some dark colouring always remains on the ears and in the majority of animals on the back also. In size it comes mid-way between a rabbit and the brown hare. Apart from the smaller size it can be differentiated from the latter by the absence of black on the top of the tail, although a negative feature such as this is not always very helpful in the field. Additional differences are shorter ears, plumper body and greyer colour. These features, taken together, should settle the matter and it is only on certain moors up to about 500 metres (1,600 ft.) that either of the two species may occur.

The next order, Rodentia, is much larger. The squirrel family has two species here and these are sometimes confused with each other. The red squirrel *Sciurus vulgaris* is the native animal. It has evolved into a distinct sub-species *leucourus* with a cream-coloured tail during the summer months. The grey squirrel from America *S. carolinensis* is of heavier build and about 8 cm (3 in.) longer. Despite this difference in size, people often confuse the two species. This is because they rely too much on colour, not realizing that there is considerable variety in both species including seasonal changes. It is not uncommon for reports to be made of the presence of a red squirrel in an area where they have long been extinct. At certain times of the year both have brownish-grey backs, the red in late winter and the grey in summer. It is obvious that colour is a most unreliable guide and the diagnostic characters to look out for in the red squirrel apart from size are the pale tail in summer and the prominent ear tufts in winter, although it must be noted that grey squirrels also have ear tufts but these are short.

To the layman it may seem a far cry from squirrels to voles but they are equally rodents although in a different family. Voles can be separated from mice by their blunt nose, dumpy appearance, short legs and relatively short tail. The bank vole *Clethrionomys glareolus* has a rich red-brown back and greyish-white belly. The body is slightly smaller than that of the field vole but the tail is longer. Young of the two species cannot be so readily distinguished and young bank voles lack the red-brown colour.

The field vole *Microtus arvensis* has a noticeably short tail about one-third of the head and body length and is generally greyish-brown in colour. The Orkney vole *M. arvalis* is the only vole occurring in these islands so that there is no possibility of confusing it with another species. As might be expected, sub-specific forms have evolved on the various individual islands. It is also found on Guernsey. This vole is larger than the field vole and of a darker colour, in fact, black specimens are fairly common in Orkney. The water vole *Arvicola amphibius*, because of its large size and aquatic habitat, is more likely to be confused with a rat than with another vole. A good view of it, however, will show the diagnostic vole features.

A mouse out in the open country is most likely to be the long-tailed field mouse *Apodemus sylvaticus* which is sometimes known as the wood mouse. It could, however, be one of two other species, the yellow-necked mouse *A. flavicollis* or the house mouse *Mus musculus*. The field mouse often has a variable yellow spot on the chest but never the broad band of the yellow-necked. This latter mammal was for long regarded merely as a sub-species of the wood mouse and in Britain it is found only in parts of southern England and Wales. The house mouse has a greyish-brown back as contrasted with the yellow-brown of the wood mouse; it also has smaller eyes, a very scaly tail and a distinctive odour. The young of the house and wood mouse are not quite so easily separated but the wood mouse has larger eyes and longer hind feet.

The most attractive member of this particular family is undoubtedly the little harvest mouse *Micromys minutus* with its bright orange-red upper parts and its vole-like appearance. Its prehensile tail, which serves as a fifth limb, gives it considerable climbing ability and enhances its appeal.

The remaining genus contains the rats of which there are two not easily distinguishable species in Britain. The ship rat *Rattus rattus* is alternatively known as the black rat but the colour is no aid in identification since some brown rats are black and vice-versa. The small points of difference are comparative ones and not of much help to the beginner. They include smaller, more slender body, longer ears, longer and

7. Wood mouse or long-tailed field mouse

thinner tail and smoother coat. The brown or common rat *R. norvegicus* needs no description being all too abundant. Slight differences in habitat of the two species are referred to in the next chapter.

The alien edible dormouse *Glis glis* could easily be mistaken by a beginner for a young grey squirrel, being similar in appearance and of the same greyish-brown colour on the back. Its very restricted range and strictly nocturnal habit should help in distinguishing them. Its total length is about 300 mm (12 in.) contrasted with the grey squirrel's 480 mm (19 in.). The native dormouse *Muscardinus avellanarius* is much smaller, only half the size of its cousin. Its body length is shorter than other small rodents with the exception of the harvest mouse from which it is distinguished, despite the similarity of colour, by the bushy tail, larger size and generally different habitat.

The South American coypu *Myocastor coypus* is one of three large aquatic mammals in this country but it should be fairly easy to separate it from the other two. Its total length is about 100 cm (40 in.) of which over a third is taken up by the tail. The colour is deep brown. Country people who do not know this mammal mistake it for a monster rat when they come across it. Some monster it would have to be, being more than twelve times heavier than a rat; their observation, however, is accurate in recognizing its membership of the rodent family. Mink and otters, the other large mammals in this environment, are carnivores belonging to the weasel family. Mink is considerably smaller than a coypu, the otter is larger and neither possesses the conspicuously blunt nose of the coypu.

The order Carnivora contains the flesh-eating mammals. The fox *Vulpes vulpes* is the only wild member of the dog family in this country; the wolf has long been extinct although as I write this there are newspaper reports of armed police searching for a pair of escaped wolves in the hills of Argyllshire so who knows whether this species may yet return to the British list. The red-brown colour of the fox is so well-known that the expression 'fox-red' has passed into the literature of colour. There is, however, some variation in colour and old animals are often markedly greyer.

The pine marten *Martes martes* is a member of the weasel family. The general colour is dark brown with creamy-yellow throat and chest. The body, while substantial, is the typical elongated shape of the weasel tribe. The tail is a long thick brush and the blackish legs are longer

CLASSIFICATION AND IDENTIFICATION

than those of its relatives. It is principally nocturnal and is not plentiful even in its main haunts. It is a tree-dwelling mammal but spends part of the time on the ground.

The stoat *Mustela erminea* is a smaller animal and much more common. It lives on the ground and is out and about in the day time as well as in the night. The colour of the back is red-brown, the under parts white but the essential diagnostic character which the observer must look for is the black tip to the tail. In northern areas during the winter stoats moult into a white coat but the black tip to the tail still remains. The weasel *M. nivalis* is a smaller edition of the stoat, similar in colouring but with a shorter tail lacking the black tip.

The polecat *M. putorius* is half-way in size between a stoat and a pine marten. It used to be called the foul marten or foumart because of the objectionable smell produced by its glands in contrast to the not unpleasant musky smell of the pine marten which in consequence was labelled the sweet mart. The facial appearance of the polecat is striking with white bands on the muzzle, behind the eyes and on the ears. The upper parts are dark brown and the under parts black. If you wish to show your familiarity with the technical jargon you will call the male a hob and the female a jill. Polecats are mainly nocturnal but they are occasionally seen during the day. Escaped ferrets will sometimes mate with polecats and some hybrids also escape from captivity so that you have to be very sure before claiming to have seen a wild polecat. The feral polecat-ferrets are usually to be separated by paler colour and less clear-cut black and white markings on the face but very dark hybrids occur. Escaped mink *M. vison* bear some resemblance to polecats and both are often found in similar watery habitats but mink are usually rather longer and heavier and they do not have the patches of white on the head.

Although the badger *Meles meles* is also a member of the weasel family it is sufficiently distinct to be placed in a sub-family of its own. The heavy grey-coated body, short legs and especially the conspicuous black and white stripes on the head make identification easy. It is rarely seen during the day and times of emergence from the sett are partly governed by proximity to dwellings; in undisturbed localities badgers may be seen well before dark. The otter *Lutra lutra* is the last of the weasels and like the badger has its own sub-family. Fortunate, indeed, will be the beginner who sees so rare and elusive an animal. Its large size, proportionately slender, sinuous body with long, thick tail which acts as a rudder in the water, makes it easy to recognize. The opportunity is another matter.

The large family of cats has only one representative in Britain and that is the wild cat *Felis sylvestris* of the Scottish Highlands. The only problem for the beginner is to be sure that he is looking at a genuine wild specimen and not one of the many feral cats or even an outsize tabby from a remote farmstead. The position is complicated by feral cats mating with wild ones so that probably much of the wild population is tainted with domestic blood. The wild cat's tail, thick and very blunt-tipped, is an important diagnostic character, separating it from the domestic animal which has a longer, pointed tail. The wild cat lacks the white spots on the feet which many domestic tabbies have. The Scottish cat is generally regarded as a sub-species *F. s. grampia* of the European wild cat.

The seals are comprised in the order Pinnipedia. Although vagrant species of seals do occur in British waters from time to time the chances of a beginner, or even an experienced naturalist for that matter, seeing one is extremely remote. So all that the mammal watcher has to do is to be able to separate the two breeding species but that is not always easy. As might be expected ability increases with experience. The common seal *Phoca vitulina* shares with the grey seal variety of colouring but the general pattern is of dark and light blotches. There are two main features which aid identification apart from the habitat differences described in the next chapter. One is the shape of the head which is rounded and in profile has a concave curve which gives it a more appealing appearance than the grey seal. The other is the shape of the nostrils which are like a V. The Atlantic grey seal *Halichoerus grypus* has quite a different shaped head and can be identified by this provided that a good view is obtained. Unfortunately, many views of seals are of distant heads bobbing in and out of the water. The crown of the head appears flattened from a frontal view and in profile there is virtually a straight line from the crown to the nose. The nostrils are wider apart than in the common seal and are almost parallel. The large bulls are distinctly darker in colour than the cows. There is, of course, a considerable difference in colour between a wet animal and a dry one.

Deer are hoofed mammals of the order Artiodactyla. They are con-

CLASSIFICATION AND IDENTIFICATION

tained within a sub-order of ruminants, animals which return food to the mouth for chewing in the process which is popularly known as chewing the cud. All species of deer are members of the family Cervidae. Red deer *Cervus elaphus* are the largest land mammals native in the British Isles. They are generally red-brown in summer changing to brown or grey-brown in winter. The rump, often known as the target, is yellowish. The males are called stags and carry massive antlers. A well-developed antler has three side points known respectively as brow, bez and trez tines and a variable number at the top which is called the crown. A twelve pointer, known as a royal, has the full complement of side tines and three on each crown. The absence of antlers does not necessarily indicate female sex since a small number of males do not develop antlers and in late winter all stags shed their antlers and are without them until new ones develop a few weeks later. Female red deer are called hinds.

Sika *C. nippon* are somewhat similar to red deer in their antlers, rather heavy build and thick neck but they are smaller-bodied, each antler has never more than four points, the coat has light spots in summer and the target is white with a black edge above. When alarmed the animal has the ability to expand the white rump in a remarkable way. The loud whistling call of the stag is diagnostic once known.

Fallow deer *Dama dama* are fairly easy to identify. The males, con-

8. Common seal swimming. Note the concave profile and rounded head

9. *Wild goats on the cliffs of Islay*

ventionally called bucks, not stags, have palmate antlers which cannot be confused with those of any other species. By palmate is meant the broadening of the end of the antler so that it is rather like an open hand. There are a number of colour varieties but the commonest has conspicuous spots on the body in summer; another variety retains these spots in winter. The rump patch is white, lined with black above and the relatively long tail is black above and white below. Females are called does and they have a more slender head, neck and body which gives them a graceful appearance. The grunting noise of the buck during the rutting season is easily recognized and carries a long way.

Roe *Capriolus capriolus* are small, attractive deer. The buck in its red-brown summer coat and with its short three-pointed antlers, is easily identified. The rump patch is a dull yellow in the buck but in the doe is white; no tail is usually visible on the buck but a short tuft of hairs can sometimes be seen on the doe. A black band round the nose and a white chin are features of the species. The grey-brown, heavier winter

CLASSIFICATION AND IDENTIFICATION

coat may make a roe doe look deceptively larger but the black muzzle and absence of tail will separate it from a fallow doe. Roe have the ability to expand the target in the same way as sika.

Muntjac *Muntiacus reevesi* are smaller still. A crouching appearance is typical caused by a rounded back and head held low. Two points are normal on the antlers of the buck. These antlers slope backwards at a greater angle than other deer. The target is white and the tail, light brown above and white below, is relatively long. The upper pair of canine teeth project from the mouth but this may not be seen in the field. It is certainly possible to confuse them with a roe but not if a good view is obtained because of the small size, general appearance, presence of tail and the angle of antlers.

Chinese water deer *Hydropotes inermis* are approximately the same size as muntjac but unlike them the bucks do not carry antlers. They have much longer tusks which can more easily be seen. The body colour is lighter than in muntjac, the ears are large and rounded and the tail is not white underneath.

Reindeer *Rangifer tarandus* exist in Britain today only as a domesticated herd in the Cairngorms and are mentioned here only for the sake of completeness. They are large animals with the interesting characteristic that both sexes bear antlers.

The last two species to be mentioned are contained in another family, the Bovidae. The appearance of the feral goat *Capra hircus* does not need a detailed description. It is smaller and shaggier than the domestic kind and the horns sometimes grow to a great length. The Soay sheep *Ovis* of St. Kilda are of two colour types, very dark and light brown and they are probably the most primitive kind of sheep in Europe. Few people are likely to see them on St. Kilda but some can now be seen in wild life parks.

CHAPTER 4

DISTRIBUTION AND HABITATS

WE must look now at the parts of the country where specific mammals live and the types of habitat which they occupy. In general, many of them are not fussy about their surroundings and can be found in a considerable variety of places. Some, however, have definite preferences although even these are not necessarily always observed by them.

Some mammals are so catholic in their choice of homes that they defy any attempt at classification in this way. The hedgehog of Britain is known somewhat loosely as the European hedgehog although there are several species on the Continent. Our animal comes from western Europe where it ranges from northern Spain to southern Norway; it is widespread in this country and is even found as far north as the Shetlands and on several of the Hebridean islands where it has been introduced. In prehistoric times it probably occupied open deciduous woodland but when New Stone Age men began cultivating grasses it can be assumed that it became a colonizer of arable land. Today it is found in most places where invertebrates are readily available as food. The number of hedgehogs seen dead on the roads in certain localities over a long period of time is an indication of high density population and possibly of a tendency to live in colonies. I used to pass one such place in a mile journey to work each morning. It was a built-up area on the outskirts of a large village with no obvious reason why it should be especially favourable for hedgehogs but over a long period it was a frequent occurrence to find one dead on the same 50 metre stretch of road. They seem to like fairly dry situations and we would not look for them in bogs or marshes. But neither are they typical of heaths and moorland although they do occur there and I have seen one on a road crossing a Shetland moor. They roost in hollow trees, bundles of wood, bales of hay, dry-stone walls and crannies in the rocks, any hole in fact which provides shelter.

Brown rats are aggressive opportunists which are likely to be found

DISTRIBUTION AND HABITATS

wherever there is anything edible and that for a rat means almost literally, anything! They especially take advantage of human activity, feeding not only on left-overs anywhere, be it under the bird table or on refuse tips, but also on main food stores from larders to corn ricks. Rats make their homes in farm buildings, hedges, sewage farms and town sewers, dwellings, warehouses, sand-dunes, docks, beaches and even in coal mines. They construct burrows in the soil, take over disused rabbit burrows, or live under floors.

The fox is another animal which eagerly seizes the opportunities which present themselves. It is primarily a dweller of the countryside, living in woods and agricultural land but it also occurs on wasteland, heaths, moors, coastal cliffs and railway enbankments. In their search for food foxes range high up on to the mountain plateaux in Scotland. In recent years they have found that they can obtain rich pickings in urban areas and reports are increasingly frequent of foxes seen in city streets as they pass on their way from one backyard dustbin to the next. They will sometimes dig their own earth but often either take over and enlarge a rabbit burrow or share a badger's sett. Foxes are well-distributed and often plentiful throughout the country but are known from only one Hebridean island, that of Skye.

Their principal prey until myxomatosis was the rabbit. These, too, were widespread in a number of habitats before this disease struck for the first time. There have been successive outbreaks but rabbits have now re-established themselves in many areas. Colonies can be found in farmland, open woodland, heaths, sand-dunes, cliffs, quarries and railway embankments. Although they burrow happily in various types of soil they appear to have a preference for dry, south-facing slopes.

Rabbits have a marked effect on the vegetation apart from the damage they do to crops. The ground around the burrows tends to become acid due to the souring of the soil from the droppings and urine. Stinging nettles and elders are typical plants associated with burrows, particularly as they are distasteful to the rabbits. But there are also a number of other flowering plants which form a recognizably resistant zone, flowers such as thistles, ground ivy, wood sage, hound's tongue and thyme. This last-named plant grows, amongst other places, on the top of anthills of the yellow ant. These anthills are the chosen sites for rabbits to deposit their droppings which fertilize the light soil and enable the acid-loving thyme to luxuriate. Some rabbits have developed the habit of living above the ground in scrub and this was possibly Nature's way of ensuring that some of the race survived myxomatosis.

Forest and agricultural land are two major habitats which are in many ways separate from each other. Each, in fact, comprises several subsidiary habitats which possess fairly distinctive plant, bird and invertebrate communities. Mammals form the exception. Whilst there are a few species which must be regarded as primarily denizens of woodland, most wander freely from pasture to arable field to copse. We have therefore to consider these areas as forming one unit.

Deer are essentially woodland creatures. Even the forbears of the red deer of the Scottish moors had their homes in the ancient pine forests of Caledonia. The destruction of most of this has forced the deer to adapt themselves to the more inhospitable moor and mountainside. It may seem strange that animals with antlers can, when required, move swiftly through the trees but this is so and those which live in forests do in general have larger antlers than those which live on moorland. The reason for this is that there are more nutrients available in wooded areas compared with the infertile moor. The Scottish animals climb to the summit plateaux in summer to try to escape the attentions of troublesome deer flies. At this time of the year the sexes are separate and often congregate in large herds. One of my outstanding memories of deer watching is of a day spent on the Great Moss on the Cairngorm plateau when I came upon a magnificent herd of between 4–500 red stags which spread out in a fan-shaped formation and moved off like a regiment of cavalry. In winter the herds move down into the glens where there is more shelter from wind and snow. Apart from the large Highland population, red deer exist on a number of Scottish islands, in Galloway, the Lake District, Staffordshire, Norfolk, Sussex, Hampshire, Wiltshire and Exmoor. Stags sometimes wander great distances and I have known them to turn up in villages where they have never been seen before and remote from known haunts. It is always well worthwhile to keep your eyes open for deer slot when walking through woods even though you may know they do not normally contain deer.

Up to the Second World War there were many deer parks surrounding mansions throughout the country and fallow deer were favourite animals enclosed in them. Inevitably over the years, occasional deer escaped and gradually feral populations were built up in the surround-

10. *A group of fallow deer. The antlers which form every year are covered with a velvety coat*

ing countryside. As a consequence fallow deer now live in a wild state in many English counties and a few Welsh and Scottish ones, including the two Hebridean islands of Islay and Mull. They have a preference for mature, deciduous woodland.

In recent years there has been a great expansion of the roe deer population in the Scottish border forests, the Highland woods and the coniferous plantations of northern England. They have not yet colonized Wales which is surprising in view of the large Forestry Commission forests there. In general, however, the spread has coincided with and is a result of the mammoth reafforestation programme of the Forestry Commission. This is yet another illustration of the fact that if you want a species to increase the prime necessity is to ensure that there is plenty of the ideal habitat available. This for the roe is principally young coniferous plantations but they also occur in other types of wood including scrub. They are fond of lying up in bogs and marshes during the day. Although small in size they are able to jump 2-metre-high fences so that there are few barriers to prevent their spread. One area each in Dorset and Sussex were the centres from which they began to spread during the last century throughout central southern England. A population has now become established in Norfolk. A small number of the larger Siberian roe live in the wild in Bedfordshire.

Sika are not such great wanderers as other deer and they occur mainly either in the vicinity of parks from which they have escaped or at places where they have been introduced. They occur in well over a dozen different areas from the Isle of Purbeck in Dorset to Caithness and Sutherland in the far north of Scotland and there are a number of herds still existing in parks. Their preference is for dense thickets and young coniferous plantations. Where their haunts abut on to reed beds flanking estuaries, as in Dorset, they will lie up in the reeds and like roe, they too will wade out into bogs.

The little barking deer or muntjac also requires thick cover and has a partiality for bramble thickets and scrub-covered waste land. For a considerable time its main centre of population has been the south Midlands but it is a great traveller and is liable to turn up anywhere. It now occurs in a number of counties in southern England but usually only one or two in any particular area and they are seldom seen due to their secretive nature and nocturnal habits. The Chinese water deer is approximately the same size as muntjac but is rarer, being found

11. The little muntjac or barking deer

in the wild only in the Chilterns and the surrounding countryside.

All species of deer move out from the woods to forage on farmland where at times they cause serious damage to crops. The view point of the exasperated farmer in these circumstances is likely to be very different from that of the nature-lover and understandably so. It may well then be necessary to take drastic action and shoot those animals which are causing the trouble. Much the best plan, however, is to take preventive action by seeing that the adjacent woodland is managed in such a way that the number of deer permitted is correlated with the amount of food supply within the wood.

Squirrels, even more than deer, are woodland animals and are largely restricted to this habitat. The branches of trees are their natural environment and they are not so much at home on the ground in large open spaces where they are much more vulnerable to predators. Even so, both species will invade gardens and farmland where their feeding activities make them a nuisance. In woodland they develop their own routes through the upper branches so that they are able to travel at speed from tree to tree. Red squirrels are particularly associated with mature coniferous plantations and the relict fragments of native pine forest still remaining in the Highlands, although they will occupy old woods containing both conifers and hardwoods. They have occasionally been seen in mountainous country in Scotland a considerable distance from the nearest wood.

Grey squirrels used to overlap with red in these mixed woodlands but in these days their areas of distribution seldom coincide since for not wholly explained reasons the two species do not flourish together. Their favourite habitats are beech woods, parkland and woods with oak standards and hazel coppice. Like the reds, grey squirrels build large nests called dreys and occasionally make use of tree holes.

Red squirrels are found today in a few places in southern England such as the Isle of Wight and Brownsea Island but their main haunts are farther north in Wales, East Anglia, parts of northern England and Scotland. Grey squirrels are widespread in Wales, throughout most of England except eastern Norfolk and parts of northern England but in Scotland they are largely restricted to the Central Lowlands.

The original homes of bats were in forests and in caves and these are their principal habitats today. Since the coming of man, however, they have adapted themselves to living in buildings—church towers, outhouses and roof spaces in dwellings. In woodland they roost in hollow trees and you may find the tell-tale marks of their droppings staining the trunk. They are gregarious animals and can usually be found in colonies although some individuals may roost singly and you may see a solitary bat patrolling its own feeding territory. Much still remains to be discovered about the distribution of British bats due to the great difficulty of identifying species when on the wing. The greater number of species inhabit southern England and they become scarcer the farther north one goes. The bats that roost in coastal caves are vulnerable to disturbance during their winter hibernation and in some places conservationists have felt obliged to place metal grids over the entrances to protect the bats.

'Brock' the badger is primarily a denizen of woods and in them most of the setts are constructed. These are dug in a variety of soils from chalk to clay but there is a preference for sand where this is available. In their nightly foraging badgers do not restrict themselves to woodland but range out over the adjacent farmland. They are often unjustly accused of killing poultry; in most cases of this sort the fox is the culprit although on rare occasions a 'rogue' badger may be responsible. A family playing in a field of corn may do a little damage here and there. Badgers are well-distributed over Britain although not so plentifully in Scotland.

The charming little dormouse is an inhabitant of deciduous woods. Its ideal environment is hazel coppice with an abundance of honeysuckle, the bark of which it shreds to provide material for its nest. Dormice are elusive creatures of rather local distribution, seldom seen unless one is accidentally discovered curled up in its winter sleep, but they may well be more common than is realized. The edible dormouse is most often evident when it occupies the roof space in houses but it is found in parkland and woods. In Britain it is restricted to the counties of Buckinghamshire and Hertfordshire.

The mole is a common animal in open deciduous woods but I suspect

12. Fresh mole heaves in a meadow

that most people if asked would immediately associate it with pasture land because there the mole-hills are very conspicuous. The density in meadows may vary from two to eight or more per acre. Moles are also numerous in arable land and will sometimes invade gardens where a solitary individual can cause havoc on lawns and borders. They are widespread throughout the country, although more abundant in some places than in others, being less plentiful on downland and in mountainous areas.

People who are out and about in the countryside during early spring will be familiar with the courtship actions of the brown hare in arable fields and meadows. It is not quite so well known that the hare is not uncommonly also found in deciduous woods. Unlike the rabbit the hare does not burrow but succeeds in making itself inconspicuous by crouching in a small depression in the grass which naturalists call a form. Although well-distributed they vary considerably in density of population, being abundant in some places, for example, Galloway and the island of Islay, but unaccountably scarce in others. In the New Forest they live on some of the heaths but are never seen on others. They find the short grass of airfields very much to their taste and at some airports they congregate in extremely large numbers.

Small mammals such as voles, mice and shrews occur both on agricultural land and in woods, some species preferring one habitat to the other. The short-tailed vole is typical of grassland where in some years it is found in enormous numbers. Its relative the bank vole is primarily a dweller in woodland and enjoys thick cover but it is quite happy living under a single line or group of trees; it can also be found along the banks of rivers where it can frequently be seen in the water, for like the larger water vole it is an excellent swimmer. Field or wood mice are ubiquitous, occurring in hedgerows, gardens, fields and woods. Some colonies of house mice establish themselves in open field situations in the vicinity of corn ricks. The delightful little harvest mouse is probably more common than is realized but it is not very often seen. The name is rather misleading for it is certainly not restricted to cornfields although a favourite winter habitat is in ricks. Reed beds, hedgerows and damp grassland are other haunts but since artists usually depict them engagingly posed on ears of wheat it is likely that the public will always associate them with corn. Of the two species of shrew, common and pygmy, the former is more numerous and occurs in a variety of situations but the latter is more local and is found in the main in woods.

Two of the carnivorous animals which prey on small mammals are the stoat and the weasel. As we travel along the roads we may occasionally see the ground-hugging form of one or the other dashing across the tarmac. But members of the weasel family go their secret ways and we must certainly count ourselves fortunate to come across them. The stoat ranges over a number of habitats but the weasel seems to prefer farmyards and hedgerows.

The mammals of buildings are very few in species. This is principally because it is not a natural habitat and any that have made man's domain their own home have had to adapt themselves to a different environment. Those that do are often subjected to a relentless campaign for their extermination which if it does not always succeed, at least effectually reduces their numbers. Bats, as we have already noted, occupy buildings because they are rather similar to their natural habitats and offer a readily available alternative. Pipistrelles and the long-eared bats are commonly found in buildings.

Rodents exploit buildings for their food potential and for the shelter they provide. The black rat is mainly confined to a few of the principal ports. It is more restricted to buildings than is the brown rat and some naturalists believe that it has a preference for the upper parts of buildings. Most people will be familiar with, and have a feeling of distaste for, brown rats which seem to be everywhere despite all men's hands being against them. House mice are sometimes joined by field mice in winter which are looking for shelter. The latter are agile climbers and have no difficulty in scaling the vertical external walls to gain access to the roof space. Once there, they make a fair amount of noise but nothing like as much as the edible dormouse which causes considerable trouble to the long-suffering householder.

Mountains and moorlands have a sparse but very select mammalian fauna. On the higher ground above 300 metres (1000 ft.) in Scotland the mountain hare begins to replace the brown hare although there is some overlap. The numbers decline above 1000 metres (3300 ft.) and within the main zone the population density varies from district to district. The main centres of population are in the Highlands, particularly in Tayside but they also occur in the Southern Uplands and have been introduced into some of the Scottish islands and in northern England. We have noted earlier that red deer live on moorland as well as in forests. Polecats are perhaps out of place in this section since they inhabit woods and farmland often at low levels but they are included here since the main occupied area is on the ancient rocks of Wales. The polecat is nocturnal and secretive, however, so it could exist in very small numbers in a few other remote localities.

The pine marten is a rare carnivore which lives on higher ground than the polecat. Its principal centre is in north-west Scotland where it is found in rocky fastnesses and coniferous plantations. It is an agile climber, perfectly at home in trees. At the turn of the century it almost became extinct in Scotland but due to the efforts of a few landowners who protected the animal, it has not only survived but in a modest way prospered. It is slowly spreading southwards and is a great traveller, occasionally appearing in areas well removed from its normal haunts. A few probably still remain in Wales and the Lake District.

Wild cats are much more widespread and more numerous than the pine marten although virtually restricted to the Scottish Highlands. They are to be found in high level woods, on rocky cliffs and open moorland. We can never go out expecting to see a mammal like this but on rare occasions one may cross the road in front of us or be found asleep on a boulder in summer sunshine or be heard purring in a crevice. These are all ways that naturalists have come across wild cats in the past and without a doubt will do so again in future. I have spent a good deal of time searching for wild cats in their known haunts without actually seeing one, although I have found evidence of their presence and that gave a quiet satisfaction. Yet when driving in lonely Glen Affric with thoughts on other things I was thrilled to see one crossing the road and I was able to watch it for a while as it crouched in a hollow at the foot of a rocky hillside and glared ferociously in the direction of the car.

When we think of freshwater, rivers, canals, ponds and lakes, we are considering a very specialized environment. The history of animals which is contained in the fossil record buried in the rocks shows that originally all life existed in aquatic habitats of one kind or another. Gradually, over many million of years, some fishes evolved into amphibious state, spending some of their time on dry land but always returning to the water to breed. Much later, some amphibians developed into reptiles and later still some of these in the course of evolution branched off and became warm-blooded mammals which lived all the time on

13. Wild cat in a tree

land. In much more recent geological time a few mammals have gone back to the water again although all of them are capable of moving about on land and some do spend a good deal of time on the ground. British mammals which fall into this category are described in the rest of the chapter.

First we will look briefly at two mammals which although they are not aquatic are nevertheless often associated with water. Daubenton's bat is frequently called the water bat because this is the bat which can be seen late on a summer evening hawking along a length of river. There are usually at least several flying together, feeding chiefly on the myriads of mayflies emerging from the water. They are capable of resting on and taking off from the water's surface. This species is well-distributed in Britain up to the Great Glen; in Scotland it is reported to be found especially in the open oak woods which abut on water, either loch or burn. Natterer's bat is another sociable species mentioned here because it usually occurs in the neighbourhood of water and is sometimes seen flying with Daubenton's. It is much more local than the latter and appears to be more plentiful in the western part of Britain; in Scotland it occurs in a few places in Strathclyde but it is most numerous in Wales.

Of the truly aquatic mammals the one which is familiar even to the casual rambler along a river bank is the ever-present water vole. It may not be known as such to him because the uninformed layman insists on calling it a water rat. It cannot be stressed too strongly that the water vole is a true vole and does not belong to the same genus or even family as the rats. The ordinary brown rat can swim quite well and is likely to be found on river banks especially where there are dwellings or when the surroundings are unhygienic with garbage scattered around. There is even a record of a rat catching fish. A report of a 'water rat', then, must be treated with caution because it could be either a brown rat or a water vole. The latter require a strongly-banked stream with deep water and a muddy bottom rich in aquatic vegetation. Wherever these conditions exist, there almost certainly you will find water voles. They make their burrows in the banks and often have an entrance below the water level. They are widespread in England, north Wales and Scotland but have not colonized the islands off the Scottish coast. We have already noted that the bank vole is not uncommon in streams and when in the water it might be mistaken for a young water vole but it is unlikely that a water vole as small as this would be out and about on its own without a parent in the close vicinity.

The largest rodent in Britain is a semi-aquatic. This is the coypu whose preferred habitat is reed swamps. These swamps abound in the Norfolk Broads, and East Anglia is the main colonizing area of this mammal. They not only burrow into river banks but also build nests amongst the reeds. They are good swimmers but are not restricted to water.

There is one aquatic mammal in the order Insectivora and that is the little water shrew. Although the largest of the shrews it is very small compared with a water vole. Unlike the vole, which is a vegetarian, the water shrew must have a stream which is rich in invertebrate life. It is distinctly local and by no means as numerous as the vole but it is scattered here and there throughout Britain. It will dig its own bur-

DISTRIBUTION AND HABITATS

rows but sometimes also makes use of abandoned homes of other creatures or utilizes a network of tree roots. It is not confined to the waterside but may be encountered a good distance from water.

There are only two aquatic carnivores in Britain and one of these is an alien. The ferocious mink is now well-established on a number of rivers. Despite heavy trapping they continue to increase and spread with rapidity. They are present on many English rivers, inhabit nearly all the Scottish counties and are found in as remote a locality as the western coast of Lewis.

By contrast, the native otter has become extremely scarce in recent years, so much so that it now receives protection. This applies only to England and Wales because it is still fairly plentiful in Scotland and is not felt to need special protection there. The habitat requirements of the otter are rivers with plenty of fish and much bankside vegetation

14. Water vole or rat feeding on waterside plant

including mature trees whose tangle of roots can provide homes which are called holts. Apart from these large holts there are smaller, temporary resting places used by travelling otters. The home range of a female is about 8 kilometres (5 miles) and of a male as much as 19 kilometres (12 miles). They sometimes travel overland much greater distances to reach other river systems and are occasionally seen far from water. As the first essential for an otter, as for any other mammal, is an adequate supply of food, this means that they have a preference in England for lowland rivers where fish are abundant; in Scotland this is not so important because even on high ground there are plenty of lochs available well-stocked with fish. Since otters have become almost extinct in England except for a small population in East Anglia the best chance of seeing signs of otters is in Scotland where it is still fairly numerous especially in western coastal regions.

We may well ask, what has been the cause of the otter's sad decline? The answer is probably a combination of several factors. Land drainage by water authorities to improve land for agriculture and the river for fishing interests often involves the destruction of shrubs and trees so that the excavators can move freely along the river banks. Weed clearance in the river itself causes disturbance. In some localities most disturbance originates from recreational activity—boating, water skiing, even people noisily walking the banks, insensitive to wild life. Pollution is another serious problem in some places, particularly in the heavily populated areas. The foregoing are probably the real reasons for the otter's scarcity rather than any natural causes although a severe winter could have an adverse effect. The result of all this detrimental activity in England is to force the otters back to the streams on higher ground where there is less food for them. It is hoped that the new conservation measures will bring some improvement in the otter's status.

Otters are not confined to freshwater. Some appear to spend a good part of their lives in the sea. Especially is this so along the western coast of Scotland and the Scottish islands.

Both common and Atlantic grey seals are essentially creatures of the sea and are very much out of their element on land, moving slowly and clumsily along the ground. They must, however, come out of the sea to breed and to moult and they love basking on half-exposed rocks or sandbanks. In the water they have the freedom of the seas and can travel long distances so that either species may be found almost any-

15. *European otter in the Norfolk Wildlife Park*

where in the British Isles. On land there is a basic distinction between the habitats although this difference is not always clear-cut. As a generalization, the Atlantic grey frequents the rocky coasts of the north and west because its required breeding habitats are rocky skerries, plateau grassland on remote northern islands, caves and sometimes sandy beaches. The common denominator in all these is absence of humans. The common seal, on the other hand, prefers inshore waters such as those parts of eastern England where there are sandbanks, and sea lochs and islands of western Scotland. It has been suggested, and this seems entirely reasonable, that the difference in habitat is related to the breeding biology. The pups of the grey seal have to remain on land for at least several days so that they need freedom from human interference. The common seal pups, however, can enter the water within a few hours of their birth, before the birthplace itself is covered by the tide. There are occasionally exceptions to this general rule, for example common seals have their pups on rocky skerries in Loch Dunvegan on Skye whilst the Atlantic grey breeds on a low sandy island (Scroby Sand) off the East Anglian coast. Seals are so scarce along the south coast of England that the arrival of one in any locality is likely to be reported as a feature of interest in the local newspapers.

CHAPTER 5

FOOD

IF we were to try and draw the food relationships of the animal kingdom in the form of a diagram we should find that it would come out in the shape of a triangle. The green material of plants is the basic food of the majority of animals so that the wide base of the triangle would consist of large numbers of invertebrates together with some birds and some quite large mammals, all feeding on plants. Above them in smaller number would be predatory insects and insect-eating birds and mammals feeding on the plant-eaters. Above these at the apex would be the relatively few flesh-eating mammals and birds of prey.

It will be seen from this that mammals are found at each level in this diagram. In addition, not only are there vegetarians and flesh-eaters but some mammals which are omnivorous and eat a wide variety of both vegetable and animal foods. Although mammals can therefore be separated into three categories of food consumers it must be understood that these divisions are not rigid and arbitrary. A vegetarian mammal, for example, is quite likely when eating leaves to swallow a number of slugs and insects inadvertently. Some carnivores, on the other hand, will deliberately vary their diet with fruit and other vegetable material.

The Vegetarians

Deer, feral goats and Soay sheep are all ruminants. That is to say, they have complicated 'innards' with a rumen which is a storage compartment into which undigested food is received. Although some digestion takes place here the material is subsequently returned to the mouth for chewing. This is the familiar process known as 'chewing the cud' which we see so often in a herd of cows lying down in a field. It is considered that this practice has a distinct biological advantage. These mammals often feed in the open where they are exposed to danger so that it is useful for them to be able to take in a large quantity of food rapidly and then to digest it at leisure in a more secluded and safer place. Wallabies indulge in partial rumination by regurgitating and remasticating some of the food swallowed. In their case the regurgitation is effected by a spiral groove in the stomach.

Most deer species both graze (that is, feed on the ground floor layer of plants especially grasses), and browse (that is, feed on tree shoots). Nevertheless their feeding habits are varied to some extent. The grass family together with the closely related sedges and rushes form the principal food of the red deer although when the animals are frequenting woodland the pattern is changed and tree shoots constitute the major part of the diet. To accomplish this they will rear up vertically on their hind legs. Bark is also stripped from tree trunks and sometimes vegetable crops in fields are attacked. Sika are mainly grazing animals feeding on a considerable variety of food but it is reported that they particularly enjoy eating the bark of hazel branches if there are hazel coppices in their haunts. Fallow and muntjac graze and browse with equal facility although their feeding emphases differ; the former are able to browse and bark mature trees which the little muntjac is unable to do.

Roe are primarily browsers feeding on conifer shoots, shrub leaves and various fruits. The food of deer generally will vary with the seasonal availability. In winter and early spring, heather, conifer shoots and other evergreens are important food items; throughout the spring and summer young deciduous foliage to some extent replaces the evergreens; in autumn fungi, acorns and chestnuts are eagerly taken.

Should you have the good fortune to come upon a shed antler which has been lying on the ground for some time you may well find that it has been extensively nibbled. This will quite likely be the work of a stag or buck itself because in this way they obtain minerals, including calcium, which are essential for antler growth. Preferred feeding times of deer are early morning and evening but they do feed at other times.

Goats have a long-established reputation for being non-selective feeders and the feral goats of coast and mountainside feed on all accessible vegetation and there is not much that is out of reach of a wild goat. The Soay sheep of St. Kilda feed mainly on bents and fescue grasses with some heather in the autumn. The red-necked wallabies feed almost entirely on heather during the winter but in summer add grass, ferns and pine shoots to their diet. They tug at the food plants and tear them off, eating with a rotating action of the jaws.

16. Red deer stag

Rabbits and hares also chew the cud but they do so by a different method from that of the ruminants. They deposit two kinds of droppings. One is moist and soft and is immediately eaten by the animal so that it passes through the stomach twice. Being eaten at once they are not observed by naturalists who only see the second type, the well-known dry pellets. This process has been named refection and its significance has only become apparent in recent years. It permits many vitamins to be absorbed by the body on the second passing through. Rabbits feed on a variety of plants including tree bark, garden and field crops but grass is their mainstay and rabbits are responsible for the close-cropped turf of chalk downs. Brown hares have similar tastes to the rabbit but they cause less economic damage. The blue mountain hare is more restricted in its diet and is especially fond of young heather and cotton grass. All three species can be seen feeding at dusk and dawn, their favourite meal times, but they also feed at intervals through the night.

Bank voles feed on a great variety of plant material—leaves, fruits, nuts, moss, roots, shrub bark, grass and flowers. A small quantity of invertebrates is also eaten. The field vole is much more limited in its choice of food, feeding mainly on grass. Water voles feed on aquatic vegetation and tree roots and have a habit of storing food in their burrows. Field mice vary their diet seasonally. For most of the year their preferred food consists of seeds but for a short while in early summer they turn to invertebrates. For the remainder of the year acorns, hazel nuts and haws are greedily consumed and also stored in underground burrows, in old bird nests and in dry stone walls. Harvest mice, as their name implies, will certainly help themselves to liberal quantities of grain but they also eat much other plant material in the form of seeds and fruits and during the summer they devour numbers of insects, too. It is possible that this animal ought not to be classified purely as a plant-eater but I am including it here with the other mice partly for convenience and partly because insufficient is known about the proportions of its food. Dormice are specialist feeders concentrating particularly on hazel nuts, acorns, fruits and berries but even these small mammals may also eat at least some insects in the summertime.

The more one looks into this question of feeding the more difficult it becomes to say with absolute certainty that any mammal is entirely vegetarian. The coypu is essentially a feeder on vegetable matter. It naturally feeds much on aquatic vegetation but also moves inland to eat young cereals and vegetable crops. There is some evidence of choosing different food items at different times of the year to obtain the best nutritive value. Yes, the coypu is overwhelmingly vegetarian but even so, when it feeds on the leaves and stems of water plants it must inevitably swallow the many small snails which abound in these situations. In addition, it is known to feed on the larger freshwater mussels found in still or slow-moving waters. It is clear, therefore, that mammals resist being placed in water-tight compartments to satisfy the tidy, orderly minds of naturalists.

Predators

We come now to the predators. Worms are the staple diet of moles at all seasons and their shallow surface runs are for feeding purposes. A variety of other invertebrates is also taken including beetle or fly larvae, ants and slugs. Occasionally the remains of amphibians and birds are found in moles' stomachs but these may well have been eaten as carrion. The consumption of wireworms, leather jackets, cockchafer larvae and fever fly larvae, all of which cause economic damage, is a beneficial activity of the mole and helps to off-set the destruction of the earthworm population. Provided that moles are not present in large numbers they probably do not constitute much of a pest on farmland. Their food varies both seasonally and according to habitat. More worms are eaten in the winter than in the summer when a greater variety of food is available. More insects are consumed in arable land than in meadows. They have been observed making soft chattering noises, presumably of pleasure, when eating. Moles store worms in their underground chambers during the winter months to safeguard themselves should the ground be frozen hard. The indigestible material which they swallow, hairs, small stones and fibrous vegetable material, form into relatively large lumps in their stomachs.

Common, pygmy and white-toothed shrews live almost entirely on invertebrate animals such as worms, beetles and wood lice which are found at ground level or in the soil. Water shrews consume small fish and amphibians as well as aquatic insects. This species has a venom-producing gland which is used to paralyse its prey which is often attacked from the rear.

Although hedgehogs belong to the same order as shrews and like them feed principally on animals they have proportionately broader grinding teeth so that they are better adapted for chewing vegetable food. They do in fact eat some plant material on occasion and their fondness for milk is well-known. Many a housewife in the suburbs puts out a saucer of milk for the nightly visit of a hedgehog which soon becomes fairly tame. A development of this enjoyment of milk is the frequent story that hedgehogs milk cows. Many scientists reject this as a fairy tale but a number of country people claim to have evidence of this and opinions differ as to its possibility. So far as I know this feat has not been scientifically proved so that the controversy continues.

Another oft-repeated story is that hedgehogs carry away fruit on their spines by rolling over on their backs. It has been alleged that this is a physical impossibility but again there are a number of first-hand accounts of this happening. The obvious moral to be drawn from this is that here is an interesting and worthwhile piece of study for you, an amateur mammalogist, to pursue. Remember that nothing is so satisfactory as observing with your own eyes. The important point to bear in mind is that you record only that which you actually see; it is only too easy to persuade yourself that you have seen that which you want to see. Another interesting fact about hedgehogs which does seem to be accepted is that they possess some resistance to poisons. They have been stung by insects such as bees and have not appeared to be in any way affected by it. They will attack and eat vipers although it has been proved in this case that this is achieved by the protection of the bristles and that should they have the misfortune to be bitten by the snake, they are susceptible to the poison and in some instances have died as a result.

Bats feed mainly on night-flying insects, especially moths, of which they consume enormous quantities. Most bats catch, eat and drink whilst on the wing. The larger species such as the greater horseshoe, noctule and serotine, prey on the large moths and on beetles such as the cockchafer whose bumbling, whirring flight at dusk is well-known to country lovers. A few species such as the horseshoes, Natterer's, long-eared and whiskered as well as taking flying insects will occasionally pick them off fences, foliage and even from the ground.

Sometimes bats are seen flying in full daylight but this will often be simply because they have been disturbed; the whiskered bat, however, is more of a day-flier than the others. This species, together with Nat-

17. Hibernating hedgehog

terer's and the horseshoe bats, includes spiders in their diet. If a bat catches an insect too large for it to eat the prey is held in a pouch or small hollow formed between the legs and the tail which is called the interfemoral membrane. Bats have poor eyesight and the whereabouts of insects is ascertained by means of the bats' marvellous system of echolocation by means of which they are able to home in on their prey. More will be said about this in the next chapter. The erratic flight of many bats is caused by their darting here and there to capture food.

The pine marten is typically a woodland animal whose ancestral prey was the red squirrel; it still is in parts of the Continent. In Britain, however, some of their habitat is mountainous country devoid of trees and even in their woodland haunts there are no squirrels present. The marten must perforce seek other game. It is agile enough to catch small woodland birds which form a large part of its diet together with small

18. *A polecat in Weyhill Wildlife Park. Polecats feed on a great variety of prey*

mammals, insects, eggs and fruit. The fruit occurring in its haunts includes the berries of the rowan trees and bilberries which are eaten assiduously.

The chief food of the stoat has always been rabbits. They are pursued by scent rather than sight and it is a well-known fact that other rabbits are passed by and indeed they themselves are not noticeably alarmed by the presence of the stoat. It is a very different matter for the pursued animal which soon becomes so paralysed with fear that it stands no chance to escape. On rare occasions, probably when a doe rabbit has young nearby, it has been known to overcome its fear. When that happens the tables are turned. I once was fortunate enough to witness on a coastal cliff in Somerset a rabbit furiously and determinedly chasing a panic-stricken stoat which had to run for its life. Since the arrival of myxomatosis the stoat has been forced to change its food to some extent and now eats more birds and small mammals. There are authenticated instances of stoats indulging in a form of play which makes birds curious and draws them within striking distance. In the winter they will sometimes hunt in small packs.

The smaller weasel can still tackle rabbits but its main sources of food are the small rodents which abound in the farmland habitats which it occupies. It will also catch birds and there are records of a bird as large as a kestrel being killed. The polecat is a larger member of the weasel family and takes similar prey in great variety although naturally it is able to tackle larger creatures than its smaller relatives are able to do. It can play havoc in the poultry runs of the Welsh farms and has been known to kill turkeys but it appears to have a great liking for rabbits and frogs when they are available. Wild cats obtain much of their food in the manner typical of their family, that is, by stalking. They feed chiefly on rabbits, hares, small mammals and carrion and are not averse to birds including poultry.

The two large carnivorous aquatic mammals are the otter and mink. The food of the former depends on the habitat it is frequenting. It travels overland and also will spend part of the year at the coast. Much of the time, however, it lives in the freshwater of rivers and lakes. Here it is distinctly unpopular with anglers because its main diet is fish. That it takes salmon and trout cannot be denied but one of its favourite fish is the eel, which is then played with before being eaten. Since eels harm fishing interests the otter is beneficial in this respect. It has an unfortunate habit of taking a bite out of a fish and then casually leaving it on the river bank, where to the irritated fisherman who finds it, it is like rubbing salt into a wound. The fish are caught without difficulty, the otter swimming underneath the fish to seize it in its jaws. Water birds such as moorhens and amphibians, particularly frogs, are occasionally caught and eaten. On land otters feed on rats and voles and one once walked into a farmhouse in search of food. They have been known to kill lambs and poultry but this is very rare. Along the shore otters not only eat sea fishes but also crustaceans from lobsters to the little sandhoppers.

Mink, like the otter, is a nocturnal feeder although it is occasionally seen swimming in the daytime. In some areas they have been reported as feeding principally on fish whilst in other places they seem to concentrate on small mammals and birds. They are fierce carnivores and liable to attack any creatures smaller than themselves. Poultry, game birds and goldfish in ponds are all at risk if mink are at large in the vicinity. Recent studies show that there is not so much overlap on the food requirements of mink and otter as was thought, although the former appear to prey mainly on land animals. There is, in fact, some evidence that mink tend to avoid otters.

Almost everyone knows that seals feed on fish. There has been much controversy in recent years over the damage done to sea fisheries including salmon netting. It has been estimated that common seals may eat about 11 lb. of food daily and the Atlantic grey about 15 lb. The fishermen allege that the seals happen to feed on the particular species which they aim to catch. The truth probably is that seals feed on a great variety of fish. Much remains to be found out about how seals catch fish in the murky depths. It is known that grey seals do not have good eyesight out of the water and even in this element it seems likely that the whiskers and nostrils are used to locate prey. They will sometimes playfully throw a fish up in the air before eating it. Apart from fish, crustaceans, shellfish and even seagulls are occasionally consumed. When seals are moulting they eat much less food and during the breeding seasons grey seals of necessity have to fast for several weeks because it is essential for them to hold their territory, there being plenty of other competitors waiting to take their place.

The Opportunist Eaters

In this category are included those which eat both vegetable and animal foods with equal relish. They belong to the 'anything goes' brigade. The fox is a carnivore but it is listed here because it is omnivorous and in particular because in recent years it has considerably changed its feeding habits. A large number of animals is certainly eaten; small rodents in quantity, large numbers of insects especially beetles, and birds including poultry. Woe betide the farmer who forgets to secure his chicken runs at night. The worst aspect is that often far more hens are killed than are eaten. The fox is an untidy killer and unlike the badger often leaves the remains of its kill scattered outside the earth. In sheep-farming areas lambs are taken so that it is not surprising that the fox is regarded as a pest in the agricultural world. Like the stoat, the fox will use a form of play to get nearer its intended victim without causing undue alarm.

Yes, the fox is a carnivore all right but nevertheless, a survey has revealed that the fox is becoming much more of a vegetarian. This no doubt is partly due to myxomatosis in the rabbit which made its preferred food very scarce. This swing in diet perhaps applies particularly to the increasing numbers of foxes which have become town-dwellers

19. Red fox in attentive posture. Note the well-developed brush

and scavengers of dustbins. Feeding is chiefly at night but it is by no means uncommon to see a foraging fox at any time of the day.

Badgers on the other hand are much more nocturnal and it is rare indeed to see one in the day time. They do not normally cause damage to agriculture by their feeding activity although a rogue animal may develop a partiality for chickens. Much of their feeding is done in the woodland surrounding their sett where they feed on a variety of plant and animal food. After dusk they can be heard snuffling amongst the trees as they set out on their hunting expeditions. A wasp nest may be excavated, a field mouse pounced upon, a litter of young rabbits dug out and a stop made under an oak to consume quantities of acorns or under a crab apple to eat the fallen fruits. Slugs, snails, roots, berries and nuts do not come amiss. Badgers also seem to be especially fond of bluebell bulbs and these are often found in the vicinity of the setts;

20. *A badger's sett in winter. The tracks in the foreground show that the badgers are still active*

since these are often constructed in light sandy soil and sandy woodland it is also a likely place to find bluebells. Their chief item of diet, however, is earthworms. They are hygienic mammals and the interior of the sett is a model of cleanliness. They dig shallow pits for their droppings at some distance from the sett and the discovery of these toilets may be the first indication of the presence of badgers.

Both species of squirrels eat varied food but in the case of the red squirrel most is of vegetable origin. Conifer seed is a favourite food and acorns, mast, hazel nuts and other deciduous seeds are also eaten. The food taken varies with the season. In spring they will take eggs from nests, eat pollen, buds and the tips of tree shoots. They follow this by eating the sap under conifer bark; then in the autumn eat berries of various kinds; bulbs and roots are dug up in the winter. Animal food consumed include insects and birds. They have a well-known habit

FOOD

of storing nuts in caches in the ground but it is not at all certain that they are able to find them again except by chance. It is considered that the cycles of plenty and scarcity of pine seed correspondingly affects the population of red squirrels.

Grey squirrels eat somewhat similar foods but are more destructive. The difference of habitat means that their depredations are concentrated on hardwoods rather than conifers. They strip bark in woods and orchards, eat corn and much fruit as well as taking young birds from the nest in the manner of a magpie. They kill small mammals and have even been known to attack and kill others of their own kind. If we are able to forget the damage that squirrels do we have to admit that the manner of their feeding, sitting upright, tail curled over back, food held in the front paws, is very attractive and never fails to please. They can, however, be quarrelsome when feeding together. Walking through a forest one autumn day I heard a tremendous din and discovered that it was coming from a a number of grey squirrels squabbling over chestnuts on the ground.

Rats are the most omnivorous of all British mammals. They eat virtually anything that is at all edible and much that is not, according to our way of thinking. Lead pipes, sewage, carrion, mammals, birds, eggs, fruit, grain, stored food of any kind, you name it and it is likely that a rat somewhere has eaten it. They usually kill by biting the neck of the victim. Like some other mammals that we have noticed earlier, rats will store food.

CHAPTER 6

BEHAVIOUR

Homes

MAMMALS find shelter in a variety of ways. One might expect the larger animals to need more than the smaller ones but this is not always so. A small proportion of British mammals have their homes below ground and one or two spend most of their time there. The largest and one of the most elaborate of these structures is that of the badger's sett. This is indeed a home in the accepted sense of the word. It has a number of entrances, passage-ways and chambers and the whole family lives in it. The sow has her young in a separate nest chamber. Setts are occupied over many years and, like human homes, are often extended. One such that I know has 17 entrances.

Badgers in general are easy-going creatures and they will often tolerate a fox coming in as a lodger. If he makes a nuisance of himself, however, he is likely to come off worst in a confrontation. Perhaps most often foxes utilize rabbit burrows although they will sometimes construct their own earth for breeding purposes. Whilst the badger carries out his major spring-clean at that time of the year the fox usually cleans his earth in the autumn. Outside of the breeding season foxes will sometimes sleep above ground provided there is dense enough vegetation available. So they will sleep in cornfields and in winter lie up in fields of kale; hedge bottoms, drains, ditches and gorse thickets are other places chosen.

Rabbits live gregariously in underground colonies. The warrens have a number of entrances and often one or two well-hidden, emergency bolt-holes. When a new warren is being excavated some holes will be sited in better positions than others and some observers believe that the more dominant and successful rabbits choose the best sites, the next in line group their holes around these while the weakest are forced to the worst position on the outskirts. This is termed a peck order, a phenomenon well-known to exist in chickens. Pregnant females often, though not always, dig a small separate burrow called a stop in which

21. *A mole emerging at the surface. The 'rowing' action of the front legs makes tunnelling easy*

to have their young. It is only these does which use bedding material. In recent years some rabbits have taken to living above ground and this seems to be a natural defensive reaction against the spread of myxomatosis which is more easily caught in the confined space below ground. They do, in any case, tend to spend more time out in the open during the summer months.

Moles not only build a complicated system of tunnels, nest chambers and the so-called fortresses but they differ from all other British mammals in finding nearly all their food underground. This is not to say that they do not occasionally come to the surface and even find some prey above ground but most of their time is spent below the surface. The disproportionately large forefeet with their great claws are a wonderful adaptation for digging and they are set at an angle outwards from the body for this purpose. The tunnels are constructed at varying depths, the deeper the tunnel the larger the mole-hills. Each tunnel system appears to be occupied by a solitary mole. Nest chambers, roughly spherical in shape, are situated along the line of a tunnel. The nest itself is made of dead leaves or grass. Where the ground water level is high, moles sometimes build what has been mistakenly termed a fortress. This contains an impressive collection of runs leading to a nest chamber. This picture certainly prevents flooding of the nest, at the same time providing it with a degree of protection.

Water shrews, water voles and coypus dig burrows in river banks for escape from danger, sleeping and breeding purposes. Certain smaller mammals such as the field mouse, common shrew, bank and field voles dig shallow burrows although they also make surface runs in litter. The breeding nest of the field mouse is often, but not always, situated in an underground chamber. It is a great climber and often uses old birds' nests for shelter although if a house is nearby it seems to prefer to spend the winter in the roof.

Certain mammals, mostly carnivores, make their homes in hollow trees, root cavities and rock crevices. Such are wild cats, pine martens, polecats, mink and otters. The last-named have two types of living quarters. The principal one is called a holt and is often situated in a bank under tree roots and usually with an underwater entrance. This also serves as the breeding den and the nest consists of grass and waterside vegetation. The other is a temporary resting place such as in a clump of the common reed or in a hollow tree and this is called a hover.

Pygmy shrews live in the ground layer of vegetation making runs in the surface litter without constructing burrows although they probably occasionally make use of existing ones made by other species. Brown hares dig shallow, open trenches which are known as forms and which would seem to give little protection but it is quite sufficient for them. Mountain hares excavate similar scrapes in peat in the summer and in the snow during the winter but they also squat in heather, dig short tunnels and hide amongst rocks. Hedgehogs build relatively large nests of grass and moss for their young and separate winter nests, in which they hibernate, situated in any suitable place such as under a shed or a pile of bean sticks. Rats and mice also construct breeding nests sited in, around and under buildings. As one might expect, rats use an incredible variety of material for their nests, some of it to our eyes, revolting.

A few mammals build nests in vegetation well above ground level. Harvest mice build nests to live in during the winter and these are

22. *A squirrel's drey in the fork of a large tree*

usually on the ground on banks or at the base of corn stacks. The rather larger breeding nests, however, are suspended between the stems of vegetation. Dormice normally build no less than three different types of nest. The winter nest, where the animal hibernates, is usually below ground but there are a few reports of old birds' nests being occupied by them. In summer they build a small nest about the size of a large apple for a daytime roost and sometimes several of these may be found in close proximity to each other. These are often situated in hazel stools or in a tangled mass of honeysuckle up to a metre or so above ground level. The nest constructed by the female to house the young is much larger, 15 cm (6 in.) in diameter and usually nearer the ground.

Of all the mammals in Britain squirrels build their homes at the highest level. These are bulky, domed structures made of twigs and lined with finer material. They are often sited near the trunk at a height of between 10 and 15 metres and are called dreys. Those of the red squirrel are smaller, neater and more compact than those of the grey squirrel.

Breeding Behaviour

When we come to look at the life history of mammals we soon realize that much behaviour is directly or indirectly related to breeding activity. This for some, begins in the cold snowy month of January. Early in this month both grey and red squirrels mate and in the latter half of the month the dog fox is on the prowl seeking a vixen, his sharp staccato barks answered by her screaming yell. It is only during this month that the vixen is in season. Rabbits have a long mating season which begins in February and lasts until September. The windy days of March see the jack and jill hares showing an exuberant interest in each other and away up on the Scottish hills the rocks echo to the fierce caterwauling of the wild tom cats. The main mating season for the majority of mammals is spring and early summer but there are exceptions to this apart from those already mentioned. The breeding season for pine martens is in late summer. For common seals it is from late July to September whilst grey seals mate at the rookeries in the autumn. Bats generally mate from the autumn through to the spring.

Courtship and aggression appear to be two opposite and incompatible types of behaviour but in some mammals they are closely associated. There are two kinds of aggression; that towards the male from a female not yet ready for his advances and that from a male directed towards other males who are potential rivals. There is much still to be learned about courtship rituals especially in the smaller mammals. Even with so common an animal as the hedgehog the actual act of mating does not appear to have been observed very often although there are written accounts of courtship behaviour. The male encountering a female circles slowly round her making snuffling noises. The female may initially raise her spines in anger and voice her objections in no uncertain manner. The male, not unduly abashed, continues his circling. If he tentatively approaches closer she uses her spines as a massed battery of spears to drive him off. The appearance presented is that of two sparring partners rather than a courting couple. Persistence usually pays off in the end, although it may take several hours.

Not all mammals are prepared to exercise that amount of patience. Some prefer the cave man technique. Bats have been observed biting the neck of the female and polecats have particularly forceful tactics for they carry the female around by this method. Shrews, too, are known to hold the female still by biting the neck. Water voles have a quite boisterous courtship which extends over several days. During most of that time the female is strongly resistant to the male's advances, attacking him with her paws whenever he comes too close. Eventually, either in the water or on the land, mating occurs.

Seals are other mammals which mate in the water; the common seal always does and the grey seal appears to prefer this although the leading bulls which have established territory at the rookery will sometimes mate on land. The bull common seals often indulge in an excited aquatic display with much splashing and many yelping cries. Two outstanding items in this display are frequent leaps out of the water and a bringing down of the fore-flipper with a resounding smack on the surface of the water. The more dominant of the grey seal bulls arrive at the breeding ground at about the same time as the cows. They carve out a territory and defend it belligerently. Quite severe wounds are sometimes inflicted upon rival bulls.

Chasing is a common part of sexual activity and is seen in such mammals as rabbits, squirrels and hares. The most spectacular behaviour is that of the last-named. All country folk must be familiar with the antics of the mad March hares. They run and jump about, sometimes leap-frogging over other jacks and frequently standing up on their hind feet to box their rivals. With squirrels several males will chase the same female and she will, as like as not, initially reject them all. Jerking of a stiffly-held tail is a typical stylized ritual of both species of squirrels. A number of rats also will chase a female and take turns to mount her although again, she may first of all show strong antipathy by kicking out at the approaching male.

During the brief courtship season in early spring, male moles become very quarrelsome and violently attack other males which intrude into their territory. A similar touchiness to the presence of other males during the breeding season is strongly exhibited in deer. A special name is given to the mating activity of deer; it is called the rut. It is a period marked by great restlessness and excited behaviour in all species.

Red stags put on a quite spectacular show. Towards the end of September tension begins to mount and they increase the time spent wallowing. This is the act of lying down and rolling on a patch of ground, a wet, peaty hollow being a favourite place. The reason for this action is little understood since it is indulged in throughout the year but at this time it probably acts as a partial release for the pent-up excitement. At this time, too, the stags begin their majestic roaring which signals the start of the rut. Each master stag gathers together a harem of hinds and the roaring is intended to intimidate other stags from trespassing upon his preserves. In October the Scottish hills echo to the challenging calls. This is not sufficient to deter other stags who try to penetrate the master stag's territory and help themselves to a hind or two. The defence of his harem is a full-time and very exhausting task for the stag. Battles with his rivals, tend, however, to be rather more in the nature of rituals rather than serious fighting. But the latter does happen sometimes and the antlers can be used with deadly effect. It is even possible that a man may be attacked if he gets embroiled with a number of fighting stags.

The roaring is not a response to the presence of rival stags but appears to be innate. I have known a solitary stag which lived for some years in a wood in southern England. Every autumn it used to leave this wood and travel two or three miles to its traditional rutting area in another wood. Unfortunately for it, there were no hinds within many miles nor stags either but for several weeks it paraded ceaselessly through the trees uttering its deep bellow and emitting the very strong, musky scent which is characteristic of the species.

23. Sika stags grazing in the Southern Uplands

The rut of sika deer takes place about the same time but instead of roaring the stag gives a peculiar penetrating call halfway between a whistle and a squeal. This call, unlike that of red deer, is given at infrequent intervals. Fallow deer draw attention to themselves during the rut by the nearly continuous grunting call of the bucks which carries a considerable distance on a still October day. This call is usually called 'groaning'.

With roe deer the time of the rutting season changes for instead of autumn it normally takes place between mid-July and mid-August. The energy which in the red stag is expressed in wallowing is exhibited in the roe buck by the fraying of tree saplings which probably serves as a marking out of territory since scent is deposited on the damaged vegetation. Territory is established some two or three months earlier. A peculiar characteristic of roe behaviour is the roe ring, the reason for which is not understood. It consists of a circle which almost always has a tree or shrub in the centre. Since it is used mainly in the rutting season and bucks have been seen chasing does round these circles it appears to be part of the rut itself although roe kids are also known to use these rings. Another obscure activity about which little is known is a false rut which takes place in the autumn. Unlike the three previous species the roebuck does not collect a harem.

Chinese water deer do not mate until the end of the year and muntjac, alone amongst British deer, apparently have no set rutting period at all.

A strange part of the breeding process in a few mammals is that which is known as delayed implantation. The fertilized egg-cells commence growing but soon come to a temporary halt for a variable period of some months. This occurs in the badger, some other members of the weasel family, seals and the roe deer. The purpose of this is not really understood. It is peculiar that stoats have delayed implantation but weasels which resemble them so closely, do not.

Some mammals have only a single litter in any one year; such are the mole, edible dormouse, fox, badger, stoat, deer and seals. Squirrels usually have two litters, one in the spring and the other towards the end of July with about three young in each. Hares produce several litters a year, usually with two leverets in each and these are placed in separate forms well apart from each other, so assisting their preservation. In favourable circumstances some mammals such as coypus, rats, mice, voles and shrews will produce a succession of litters throughout the year.

Single young is usual in seals and most kinds of deer but a large percentage of roe births produce twins and Chinese water deer have up to four fawns. It should be mentioned at this point that female deer leave their young lying down in a safe place while they go off to feed. These fawns are not neglected by their mothers who know exactly where they are. Every year there are newspaper reports of people with kindly intentions but totally misguided actions who rescue 'abandoned' fawns from the depth of the woods. Not only is unnecessary distress caused to the parent but the rescuers will have some difficulty in bringing up the youngster and even if they succeed they will only have to release it in the wild where its tame ways will probably prevent it from surviving for long. Fawns discovered in the undergrowth should be left there and should not even be touched.

BEHAVIOUR

Play

The young of some species indulge in play. Young squirrels will chase each other and generally have fun and games. They will bury objects although it is doubtful whether they have seen their mother doing this. Fox cubs are always attractive animals to watch as they leap about and practise pouncing upon one another at the entrance to the earth. It is said that it is usually only in May when they are out in daylight hours. The vixen leaves them when they are only six weeks old. As they get older they naturally become more active and make more noise during their play. Young badgers can be very noisy too and I have heard them yelping well before I reached the sett. Wild cat kittens play in the manner of domestic kittens outside their lair from four weeks old onwards but there must be few naturalists who have been fortunate enough to watch such an entrancing sight. A number of young stoats will gather together for boisterous play with mock fighting, jumping and somersaulting. Otters are naturally playful animals, both young and adults. One of their favourite pastimes is for the family to slide down a snow or ice-covered bank, repeating the process over and over again. Amongst the small mammals the young of the lesser white-toothed shrew have a habit which to us looks like play but which no doubt is very serious to them. They accompany their mother on her excursions and to save themselves getting lost the first youngster grasps hold of the base of its mother's tail, the next one grasps the tail of the first and so on until they look like a succession of trailers. This process is called 'caravanning' and is a very amusing sight to watch.

Social Relationships

In general, outside of the breeding season, individual wild mammals go about their own business taking little if any notice of other mammals. There are, however, some notable exceptions to this. The predators obviously take a special interest in their prey! Then there is a certain amount of family life in some species. Badgers are typical of an extended family existence in that more than one family will sometimes live amicably together in the same sett. They are also known to pay occasional social calls to neighbouring setts.

For most of the year roe deer live together in small family groups of buck, doe and young. This breaks up in the spring when the doe drives the yearlings away and the current year's kids are born. After the rut at the end of July the buck wanders away for a few weeks before rejoining the family. In severe winter weather a number of roe will band together but one does not normally expect to see more than three or four roe together. Other species of deer do not have family life like this although muntjac pairs appear to live in the same territory making contact from time to time by means of scent-marking. Mixed herds of fallow deer occur in winter but throughout the summer the sexes keep separate from each other. Generally these groups are not very large, consisting of up to 15 or 20 animals. In the case of red and sika deer the sexes are usually apart except during the rut.

Although hares are often seen singly both British species will occasionally meet in larger numbers. This is especially so in the case of the mountain hare which when driven from the high ground by snowstorms will gather together on the lower slopes. Rabbits, of course, have a communal life and their habit of thumping the ground with their hind feet

24. *Tree showing the characteristic fraying by roe deer*

to warn of danger serves the community well. Rats are prolific breeders and a pair finding a congenial unoccupied haunt soon grows into a large colony with a tribal loyalty which may motivate them to attack other rats which seek to join the resident population. Seals often live together in a loose association outside of the breeding season when they move to feeding grounds well away from the rookeries. They can be seen hauled out on sandbanks and rocks, and they seem to need and enjoy these periods of basking between their fishing excursions.

In the autumn some unusual mass movements of squirrels have been observed, the reasons for which are not entirely understood. These can be partly explained by young dispersing from breeding areas but since many adults are also involved this is clearly not the entire answer. The hunting in packs of weasels and stoats has been mentioned in the previous chapter. Most British bats crowd together in summer roosts and winter hibernating quarters. These colonies may number several hundred individuals. Pipistrelles are able to gain access to roosting areas through astonishingly small crannies as little as one centimetre in width. In the case of this bat and some other species the sexes are generally segregated in the summer roosts.

Calls

Many mammals are surprisingly silent for much of the time. One which the householder may well hear on a summer night is the hedgehog. Unless he has heard it before he will quite likely be puzzled by the sounds, which vary from low grunts to high-pitched squeals. Two animals fighting can produce some disproportionately loud squealing. Camping on the Somerset coast, my family and I were woken up in the middle of the night by a medley of sounds coming from the outer tent. With hearts thudding furiously we peered out only to see a hedgehog busily devouring the contents of a saucepan! Shrews make frequent soft squeaks as they run about but very good hearing is required to pick up these sounds.

If good hearing is needed for shrews, it is even more essential for the sounds emitted by bats. In fact, some calls are only likely to be heard by young people. The audible calls of various bats differ to some extent but can be described either as a sqeaking or a chattering. In addition to these audible sounds bats possess the extraordinary ability to navigate by making ultrasonic sounds. Their eyesight is so poor that they would never be able to avoid obstacles while flying unless they had some means of locating these obstacles. Nearly 200 years ago it was discovered that bats whose hearing had been tampered with, bumped into objects. Unfortunately no further research was done until the present century when reflected ultrasonic sounds, a kind of natural radar, was first suspected and later, conclusively proved. What happens is this. The bat emits an almost continuous series of inaudible sounds which are reflected back from any object in the route ahead. The time taken for the echo to come back enables the bat to assess how far away is the obstacle. How it distinguishes edible prey from other objects remains a mystery. Zoologists have given the name 'echo-location' to the bats' method of finding their way about.

Rabbits and hares are exceptionally silent animals. Both species, however, will grunt when excited and scream when attacked. Squirrels are much more vociferous, making soft conversational calls to each other and also loud chattering cries when annoyed. Voles and mice squeak and chatter, some species more than others.

Carnivores make a variety of calls but often for much of the time are silent. The typical call of a fox is a bark but a loud, penetrating scream is also quite common. It used to be thought that the scream was produced only by the vixen but it is now believed to be made on occasion by the dog fox as well. Badgers give warning barks and they have various other calls including a scream not unlike that of a fox. These screams of fox and badger can be quite frightening if you are alone in a wood at night.

Most of the weasel tribe make chattering calls which are similar to each other. What will certainly bring the females into voice is if there is any threat, real or imagined, to their young; they will then erupt into angry, staccato ejaculations. When alarmed, polecats and mink will shriek but otters give an angry growl. The characteristic note of the otter, however, is a clear, piercing whistle which helps to locate the animal. Wild cats make the usual feline sounds of contented purring, mewing, growling and the fierce caterwauling of the toms. A Scottish naturalist once had the unusual experience of hearing the rhythmic snoring of a wild cat coming from a hollow tree stump.

The two species of seals differ considerably in their vocal activity. The common seal is often silent but yelping, wailing cries are heard during the breeding season. Grey seals frequently make a variety of

25. Common seals basking on a Hebridean skerry

sounds. They hiss at intruders but the most well-known sound is the so-called seal singing, a quavering, moaning hoot, once heard never forgotten, a sound which seems so perfectly in harmony with the turbulent seas which are their home.

The distinctive rutting calls of red, sika and fallow deer have already been mentioned. Outside of the breeding season red deer are usually silent, particularly the hinds, although even these will give a warning bark occasionally especially to their calves. Both hinds and calves use a bleating call when making contact, or trying to make contact, with each other. Sika make short squeals or screams when alarmed. Fallow bucks and does will sometimes give warning barks and the does and fawns communicate in the same way as the other species. Muntjac, sometimes known as the barking deer, have a staccato bark but they also occasionally make a repetitive clicking sound which is quite distinctive and has been likened to the rattling of castanets.

Coat

Mammals are distinguished amongst other things by the possession of a body-covering of hairs. This coat or pelage is important for several reasons, protection from cold, camouflage and even defence in the case of the hedgehog. Instead of hairs the hedgehog has one millimetre thick striped spines on the upper surface of its body. When the animal is in a passive state the spines are laid flat but when alarmed they are held erect and the hedgehog rolls itself into a ball to become a fortress which can successfully hold some attackers at bay. One 'attacker' which it fails to hold back is the motor car which is responsible for the death of thousands each year. It is believed by some naturalists that a different habit-pattern is emerging and some hedgehogs are now ceasing to squat when danger approaches on the road. This of course would have great survival value. They have a strange custom of anointing their spines with saliva; it appears to be stimulated by the presence of various scents but the reason for this practice is not known.

Most mammals moult, that is, renew their coat, at least once a year. Annual moults are seen in the rabbit, fox, badger, otter and seal. When the otter and other aquatic mammals swim their progress is indicated by a line of bubbles caused by air trapped in the fur. The two species of seal have different moulting times and in the grey seal even the sexes differ. Common seals change their coat in late summer, grey seal cows in the first three months of the year followed by the bulls from March to May. Young grey seal pups have a woolly coat consisting of long white fur which is shed during the fourth week.

Twice-yearly moults are common and mammals in which they occur include shrews, brown hare, squirrels, stoat, weasel and deer. An interesting fact of these moults is that in the spring the hair is renewed from the head first proceeding towards the hindquarters whereas in the autumn the procedure is reversed. In the autumn the coat which develops is thicker than the summer coat, often of a different colour and with dense under-fur. An unusual phenomenon which occurs in red squirrels is that in addition to the moults successive changes in colour take place during the winter due to chemical alterations of pigment in the hairs themselves. The winter coat of the stoat is usually white in northern areas of Britain where much snowfall occurs. This is a striking example of a habit evolving which through camouflage has survival value.

BEHAVIOUR

The mountain hare is unusual in that it has no less than three moults each year. The second moult follows only a few weeks after the spring one; in the autumn the third moult takes place producing a white winter coat as protective coloration.

The casting of the antlers of deer is in some ways akin to the moulting of the coat. Usually this is an annual event. Antlers grow from bony lumps called pedicles which develop on top of the head. These appear within a few months and are followed in the second year by immature antlers in the nature of spikes. In succeeding years the typical antler for each species is grown, the number of points generally increasing with age up to a peak, although dependent to some extent on food and climate, and thereafter declining as the body itself deteriorates. Each year the new antlers are covered with a skin called velvet which dies and is rubbed off. Red and sika stags cast their old antlers about March and by August the new antlers are clean. Fallow bucks cast theirs rather later, in April and May, but the new ones are clean by August. Roe drop theirs during November or December and are clean by the following May. An occasional roe buck may be seen with antlers permanently covered with velvet; bucks in this condition are called rather fancifully, perruques which is an alternative name for a wig.

CHAPTER 7
HOW TO OBSERVE

WE have already seen in the first chapter that looking at mammals is an occupation to test all our skill and ingenuity. This, of course, is true to some extent of every branch of nature study but never more so than in the observation of mammals. It comes down to a kind of detective work where much of the enjoyment derives from patient searching for clues and where in a few cases the body may never be sighted at all. But as in crime detection, patience, persistence and skill are likely to bring rich rewards.

It is essential to know the habits of the mammal we are looking for; whether it is active by day or night; what type of habitat it frequents; its preferred food; what kind of home it has and what its behaviour patterns are. Much of this groundwork has been covered in previous chapters and now we come to the interesting and at times exciting practical work of looking out for the tracks and other signs of mammals and occasionally watching the animals themselves.

Certain principles apply in the search for all mammals. Try and walk as the larger mammals do when they are foraging, that is, slowly but in an alert manner, walking with as light a step as possible and pausing after every few paces. Be prepared to 'freeze' instantly on seeing any suspicious movement. Camouflaged jackets or at least subdued clothing should be worn at all times and where this can be done, walk into the wind, certainly never have the wind behind you when looking for the larger animals.

Another technique is to settle yourself down at a good vantage point overlooking a fairly open stretch of land where you have reason to believe that mammals may appear. It may be that it is a field which deer are known to frequent during the course of an evening or there may be recently-used wallows or even a roe deer ring. It may overlook a pond where animals come to drink or a fox's earth where cubs are believed to be present. This method may well be productive but it requires a great deal of patience and on a summer evening liberal doses

HOW TO OBSERVE

of insect repellent will probably be needed. In woodland listen carefully for sounds of movement. Blackbirds probing about in leaf litter will initially mislead you but after a while you will learn to distinguish between them and scurrying grey squirrels.

Tracks

The most conspicuous evidence of the presence of mammals is provided by their tracks. A good lookout should therefore be kept for soft or wet ground where footprints will show up clearly. If you want to know if a particular wood contains deer or not, then a check for tracks will give the answer.

Tracking has its own specialist vocabulary and it is helpful to learn the few terms which are involved. Slot is the individual footprint. Trail is the track or succession of footprints. An imaginary central line running between the pairs of prints is called the median line. When the hind feet are placed on the slot of the forefeet this is known as registering. The feet of deer and similar animals are cloven, that is, they are divided into two halves which are called cleaves. In soft ground the cleaves spread out and this is termed splaying. The track of an animal through grass is known as foil and if the animal had been coming towards you the track will look darker than the surrounding grass. Conversely, if you are following the mammal the trail will look lighter. Deer have two vestigial toes at the rear of the foot which are known as dew claws. As they are higher up the leg they do not normally make a print but in certain circumstances such as in soft snow, their impressions may be seen on the ground. Plantigrade animals are flat-footed, treading on the rear part, unlike the majority of mammals which carry their body on the toes. Badgers, otters and hedgehogs are examples of plantigrade mammals.

We will look now at the tracks of a few common species that the beginner is most likely to come across; to deal with all would require a whole book by itself. In woodland the most frequent and conspicuous tracks to be seen are those of deer. We have already noted that deer have distinctive cloven hoofs; the only other wild mammals with similar prints are wild goats but these in the main occupy different habitats. Experts may be able to distinguish the sex and even the age of the animal but we shall be more than content for a start just to identify the slots as being those of deer. After a little while we shall gain experience

Fig. 2. Footprints of a number of well-known mammals and rabbit trail

Squirrel print of fore foot

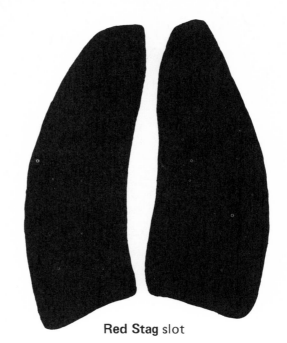

Red Stag slot

Slot of **Roe Buck** galloping *Note splayed cleaves and dew claws in evidence.*

Hedgehog print of hind foot

Stoat print

and be able to separate some if not all of the different kinds of deer. The identification will be largely based on size of slot although we must take into account the fact that prints of males will be proportionately larger than females of the same species.

Another important point to bear in mind is that the size will vary according to the nature of the ground and the speed of the animal. The largest slot is made by red stags and is about 8 cm (3 in.) in length. Next in size is that of fallow which is not only rather smaller but is proportionately longer. The slot of sika is slightly smaller still but the beginner cannot expect to separate them from fallow and will have to rely on other evidence such as information from local keepers or a direct sighting of the animal itself. Roe tracks however will become readily identifiable with a little experience. They are distinctly smaller than those of the three preceding species, the cleaves are relatively narrow and as they have a tendency to spread the effect produced is of a small, squarish print. If you should be so fortunate as to find some tiny slot in which the cleaves are unequal in size you will know that you have stumbled on evidence of the presence of muntjac although these impressions are sometimes symmetrical.

So far, all this is quite straightforward but a complication now arises. Deer are not the only hoofed mammals roaming over the countryside; there are all the domestic animals of the farm. Sometimes the habitat will be a guide. For example, we should not expect to find cattle or sheep in a fenced lowland plantation; on upland farms, however, an open wood may well have tracks of both deer and sheep.

Deer and domestic animals can be differentiated by several points, each one comparatively slight in itself but taken together they give a cumulative picture which is usually sufficient for identification. Of course it must be understood that there is quite a variety of domestic breeds and this in itself means that the size of prints will vary. In general, however, cattle tracks are approximately the same size as those of red deer and those of sheep and goats are similar to the slot of roe. The prints of domestic animals differ from those of deer in being broader, having more rounded tips and more distinct heels. They may appear larger than they are because the feet are often dragged. The prints do not lie so close to the median line as in deer.

Another set of tracks which will commonly be found is that of the fox. Amongst gamekeepers the tracks are sometimes known as 'pad-

26. *Roe deer slots in clayey soil*

ding'. The prints show four toes and a quite small pad. The size of the tracks is about the same as that of a small dog such as a terrier so that you may wonder how they can be distinguished. There are several small points of difference which although slight in themselves, taken together usually enable a determination to be made. As with many wild animals the fox treads mainly on the toes so that the heel pad is not so distinct as in the case of a dog. In the fox the prints lie close to the median line whereas in a dog they tend to be further away. When a fox is walking or even trotting, the hind prints register on the fore. The heel pad is approximately the same size as the toes in contrast to the larger pad of the dog. The claws of a fox are finely pointed and usually only in evidence when the animal is galloping. In a fox the front toes are more widely separated from the hind ones than in a dog. Dogs also tend to run about erratically whereas a fox gives the impres-

27. *Rabbit footprints in snow*

sion that it knows where it is going and frequently-used routes are the headlands abutting on to hedges.

When walking on clayey woodland paths you should always look carefully for badger tracks; they are quite different from those of fox. For a start, they are not so elongated as those of a fox but are relatively square in outline. Badgers have five toes although this is not a very reliable guide since the print of the innermost toe is by no means always visible. The most distinctive feature is the very broad and massive heel pad.

Rabbits and hares have similar prints to each other, the chief difference is that those of the hare are larger and are spaced more widely apart. The trail shows a highly distinctive pattern of two forefeet almost in a straight line one behind the other and the much larger hind feet which are ahead of the forefeet and spaced out on each side of the median line. Gamekeepers sometimes use the term 'prickings' to describe the trail of the hare.

Squirrels have four toes on the forefeet and five on the hind. The long toes which usually show claw marks and the divided heel pad, help to identify the tracks. Red squirrel tracks often do not show the marks of the heel in contrast to those of the grey squirrel which are normally in evidence. Squirrel trails are distinctive with hind feet in front of the fore and splayed out farther from the median line.

Animals of the weasel tribe, stoats, weasels and polecats, have five toes on both fore and hind feet. Due to the light weight of the body the tracks are often not clearly impressed. The normal gait is a bounding one with sets of four prints but when these animals are hunting the hind feet register and consequently the prints are in sets of two.

Like the weasels, hedgehogs, too, have five toes on all their feet. The toes are short and blunt in contrast to the long claws. Their usual method of movement is a run and the trail shows sets of two prints alternating on each side of the median line. These are the common, larger mammals whose tracks you are likely to encounter. When you have become familiar with these it may be that your interest will have been aroused and you may decide to take up tracking as a hobby. If so, you will need to acquire one of several specialized books dealing with this fascinating subject. One final word of warning; do not expect to find perfect tracks. You may have to search along a trail to find prints which are reasonably good.

There are signs other than footprints which the observer can look out for. Where deer emerge from dense cover on to more open ground they make a path which is known as a rack. The ground on a bank is often worn bare. Badgers make narrow paths through vegetation. Mountain hares make distinct paths through heather which often run for a considerable distance. Where rats run along rafters or walls they often leave typical smears where deposits of dirt have been left from their coat or paws. In East Anglia a devastated area in a reed bed, where the vegetation looks stunted and sick, is likely to be the work of coypus.

Scent

Some mammals advertise their presence by depositing scent at certain places in their territory. This action probably serves a dual purpose; it enables a pair to keep in touch with each other and it also warns off other members of the same species. The scent of the fox is perhaps the one you are most likely to sniff in a country walk. Even if you do not possess much of a sense of smell you should have no difficulty with this. The problem may be in identifying the smell since it is not possible to describe a smell except by reference to some other scent which is known. We call it a foxy or musky scent but that does not get us very far. It is best if you have a companion who is familiar with the smell; once identified it will never be forgotten.

Another carnivore with a powerful scent is the polecat but this is an exceedingly unpleasant one. The foul scent is especially noticeable when the animal is annoyed or frightened. With the exception of the seals, all the carnivores possess scent glands but the two mentioned are probably those whose scent is most likely to come to the notice of the observer provided of course that he is in their particular haunts. In the rutting season, as we have seen earlier red stags emit a strong scent which lingers on the ground. In other species of deer it is less noticeable.

Droppings

An even more valuable guide to the existence of a particular species of mammal in any locality is the presence of droppings. You may feel that it is the animal itself which you wish to see and may be too impatient to look for signs. That would be a great pity. To long to see the mammal itself is understandable but the preliminary work can seldom be short-circuited. After all, there is little point in spending many days searching a wood for a mammal that is not, and never has been there. My own garden has been visited by fox, rabbit, mole, rat, field mouse, bank vole and common shrew but of these seven species I have only actually seen alive there, individuals of three species.

Deer droppings occur in clusters. In mediaeval hunting language an individual deer dropping was called a fewmet and the heaps were named crotties. In recent years there has been an attempt to revive the use of these specialized terms, without much success. These droppings are oval in shape and those of red and fallow are often adherent. With experience, the droppings of these two species can usually be distinguished but you will have to be guided by the size. If the individual droppings are 2 cm ($\frac{4}{5}$ in.) or more long they will be those of red deer, if between 12 and 20 mm ($\frac{2}{5}$ to $\frac{4}{5}$ in.) they will probably be either of fallow

28. *Fallow deer droppings*

or sika. Those of roe are smaller still, under 12 mm, usually nonadherent and sometimes scattered along the ground.

Fox droppings can frequently be found. They resemble those of a dog but differ from them in being generally of a smaller size, containing much hair and having one end twisted into a point. They also often conspicuously show the wing-cases of beetles together with the bones of small mammals and birds. If you come across some shallow pits containing dog-like droppings rather larger than those of a fox but similarly often containing insect remains you will know that you have come upon a badger's latrine.

It is sometimes believed that wild cats, like their domestic kinds, are scrupulously careful not only to excavate a hole but also to cover their faeces. This is only partly true. I have more than once discovered wild cat droppings on path and rock with no attempt made to cover it up.

On some occasions, however, they are known to bury them. It is thought that the truth may well be that they are left exposed on the boundaries of the territory in order to act as a scent-marker. Certainly these faeces have the same offensive smell as those of the domestic animal.

The droppings of otter by contrast have a not unpleasant, almost sweet smell, at least when they are old. When they are fresh there is a more distinctly fishy smell. They are usually black in colour when fresh, grey when old. These droppings are called spraints and they are regularly deposited on certain conspicuous spots such as a rock in midstream, a tree stump and similar places to mark out and lay claim to its territory.

Rabbit droppings are frequently deposited on ant hills and this must also be for a territorial purpose. Hare droppings are of a similar fibrous nature but larger, being about 15 mm ($\frac{1}{2}$ in.) in diameter. Bat droppings resemble the spindle-shaped ones of mice but if they are dissected under a pocket lens they will be seen to consist of many insect remains. Small mammal droppings usually contain hairs of the animal itself and scientists can determine the species by examining these hairs under a microscope.

Food remains

So far we have seen that evidence of mammals can be gathered from tracks, paths, scent and droppings. This does not exhaust all the possibilities. The remains of food scattered on the ground will sometimes give a clue as to the mammal responsible. The shells of hazel nuts are always worth examining. If you find shells which have been split lengthwise this will be the work of a squirrel. Whole shells which have an untidy hole showing toothmarks both on the inside and outside are indicative of the feeding of a field mouse. Holes with toothmarks only on the inside are probably made by bank voles and neat holes with toothmarks only on the outside are produced by dormice. Creatures other than mammals will also feed on hazel nuts and it is helpful to be able to recognize their handiwork. A tiny, extremely neat hole is the work of a beetle, the nut weevil. Nuts seen in a crack in the trunk of a tree are likely to have been placed there by a nuthatch or one of the woodpeckers.

Pine cones largely stripped of scales except for a little tuft at one end,

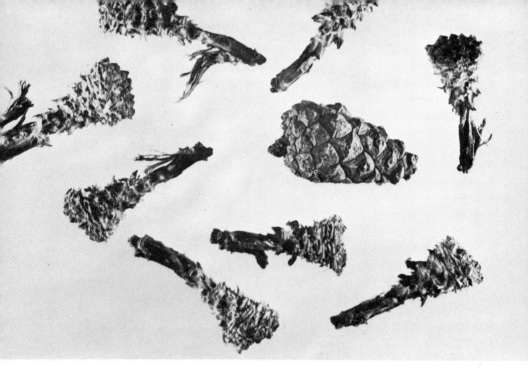

29. *Pine cones stripped by grey squirrels*

show that squirrels have been feeding on them. A store of berries in an old nest is certainly the hoard of a field mouse and these stores may also be found on the ground under boards although here it could be some other small mammal.

Particular Techniques

If you are fortunate enough to find a hedgehog in the garden be sure to put out food to hold its interest in the locality; left-over scraps from meals will be acceptable. They are very partial to slugs and any gardener would be anxious to entice and keep them on this account alone. I once knew of a dog which had a habit of collecting hedgehogs and bringing them unharmed to its owner.

Moles are very difficult to observe since much of their life is spent underground. If you want to examine one it is sometimes possible to see a run being excavated by a mole just below the surface; quick action with a spade or similar tool may well enable you to throw the mole out especially if at the same time a foot is placed to block its retreat. Another way is to use a scissors trap placed in a mole run but this of course kills the animal. Scientists can trace the movements of moles by attaching a radioactive substance to the base of their tails and then following their journeys by using a Geiger counter but this method is definitely not available for beginners!

Much searching of suitable habitats during twilight hours will be necessary to locate colonies of bats. Explore open woodland, tree-lined lanes and river banks. Remember that many bats do not appear until sunset or later so do not be in a hurry to give up the search; in fact on a moonlit night you could with advantage carry on watching for a long time. An alternative method in the day is to search a wood for tree trunks which show staining of bat droppings below a cavity. This can then be followed up by checking at dusk to see if bats are actually in residence.

A hide erected on the leeward side of an extensive warren could provide absorbing hours watching rabbit behaviour. I once spent a holiday in a beach bungalow on dunes which were swarming with rabbits and they could be watched in comfort from any of the windows. Hares present a different problem altogether. Early spring is the time when they are likely to be so preoccupied with their antics that they allow a closer approach than is possible at other times. A boundary hedge will often give some cover and you can then wait in the hope that they will move nearer. At any time of the year resting hares can with experience be picked out in open country by their typical rounded shape; after a bit of practice the eye becomes adept at picking them up and the identification can be confirmed with binoculars.

Since water voles are creatures of the day they are easy to see provided that you keep still. Choose a spot where burrows are present in the river bank and then sit down and wait. They make a distinctive 'plop' when entering the water and this can alert you when the animal itself may be hidden by overhanging vegetation. They can also be heard gnawing at the stems of plants. It is reported that they are attracted by the smell of apples and it would be worthwhile experimenting by throwing small pieces on nearby mud where voles can be expected to land.

HOW TO OBSERVE

Foxes wander over a wide area and most naturalists would have to admit that nearly all their encounters with this mammal were chance ones. There are ways that you can increase your chances of seeing one if you are really determined. Foxes pair up in the first few weeks of the year so you can go out on moonlit nights, listen for their calls, and try to track them down. Later you can visit all likely spots to see if you can locate an earth where the vixen has cubs. These places should have been found in preceding months with the help, if possible, of a friendly gamekeeper. If you find an occupied earth indicated by the animal remains outside do not get very close or you will spoil your chances because the vixen will move the cubs to another place; 50 metres might be a good distance. Joining the foot or car followers on a winter's fox hunt is another possible way of glimpsing this animal.

Badger setts, being so much larger, are easier to find. On a cold winter morning you may be able to see vapour rising at the entrance; this is a sure indication that badgers are in residence. If not you must look for other signs. Have the paths been used recently? Are there fresh claw marks on nearby trees? Can you find recently-used latrines in the vicinity? Are there any signs that bedding has been taken into the sett? If any of these answers are positive you can confidently arrange to come back at night. If you have walked all over the sett, however, you must understand that your scent will linger for several hours and it may be advisable to defer the visit until the following night.

Before you go, take a last look round at the surroundings. See if you can find several alternative positions for watching a particular hole so that whatever direction the wind is, you will have an appropriate watching point. With well-camouflaged clothes it may be better to stand in front of a tree rather than behind it. If you are behind a tree you have to keep peering round the trunk and the frequent movement and unnatural shape of the tree will make the badgers suspicious. If you can get above the ground in a low fork so much the better provided that you are actually above the level of the entrance.

You can never be certain when badgers are going to emerge. It is therefore advisable to be in position a good hour before you are expecting to see them. This is when all your patience will be called into play; an hour seems a long time when you are standing absolutely motionless. At the risk of repetition I repeat the word, patience. On my very first visit to watch badgers I had two cubs running up to my feet and

30. *Pine marten scrambling on rocks*

although I was excited by this experience I did not then realize how lucky I was. On many subsequent occasions I have waited in vain without catching a single glimpse of a badger. In time, however, perseverance will bring its own reward.

When a particular animal is restricted to a relatively small area of the country there is often the feeling that if only we could spend the next holiday in that region we would surely be able to see the mammal. Nothing is further from the truth. Such creatures are often thin on the ground in their own territory and elusive in the bargain. That is no reason, however, why an attempt should not be made. After all, as any hunter will tell you, much of the pleasure lies in the hunt itself regardless of the outcome.

I once spent an entire holiday in looking for pine martens. Yes, you guess right, I never saw one but it was one of my most enjoyable holidays, nevertheless. Looking back on it, I do not think there was much more that I could have done but the odds against a sighting were too many. In such circumstances we have to be satisfied with obtaining evidence of its presence.

I did meet people who had had interesting encounters with martens. One couple had had a marten visiting a lump of fat hung on a branch for the birds. One night it had seen its own reflection in the house window and had shadow-boxed with it. Then an old man in a lonely cottage told me that a year earlier he had seen a female leading a procession of young ones, a most attractive sight. For myself, despite an all-night vigil in the depths of a remote forest, I had to be content with finding the typical marten droppings.

Mammals of the weasel family to which the marten belongs are almost always seen by chance. They have, however, one trait which we can turn to our advantage and that is their innate curiosity. If you see a stoat or weasel disappearing into the crevice of a stone wall it is always worth while stepping back a short distance and then waiting patiently. After a little while you may well see an inquisitive head peeping out. If you are adept at mimicry you may be able to attract a stoat or weasel by imitating the squeal of a frightened rabbit. It is also reported that stoats can sometimes be drawn out of hiding by the waving of a large feather.

Deer are not seen as often as one might expect considering their large size and the fact that they often roam in small groups. They have acute hearing, a keen sense of smell, good eyesight and are skilled at concealment. Nevertheless, they are animals which a beginner is probably most likely to see if he proceeds carefully. I climbed a Scottish hillside one day with a small party of people led by a local naturalist. All around us were ranges of heather-clad hills which to us seemed at first devoid of life except for a few meadow pipits. The practised eyes of the local man, however, were picking up red deer, one after the other, on the ground high above us. It was simply a question of sharp eyesight and experience; our eyes were just not registering the small lumps which to him stood out clearly.

In the case of a herd of red deer in hill country an old hind usually acts as a lookout and it is therefore helpful to try and pick out this animal when a long distance away. By keeping your eyes on her you can freeze the moment she looks in your direction and not proceed until her suspicions are allayed. In this type of terrain look out for peaty hollows where slots indicate that deer have recently indulged in wallowing there. Always move very cautiously when emerging from a sheltered area to a more open one such as when walking round the base of a hill.

Similarly, in woodland when you come to a cross-rides you should always pause and look very carefully each way in case there is a deer feeding along the side of the ride. Roe in particular are fond at dusk of feeding in ditches where they are inconspicuous along the edge of tracks. Some people are able to attract sika towards them by placing a blade of grass on the lips and whistling through it. Fallow are perhaps the easiest of all deer species to see. They are active by day, although their main feeding times are in the early morning and evening.

It is an excellent plan to climb up into a tree on the outskirts of a large glade frequented by deer. Animals seldom look up and often it prevents the deer from 'winding' the mammal-watcher although it cannot be safely assumed that currents of air will not sometimes swirl around you and carry your scent to the animal below.

Although looking at mammals can be accomplished without any equipment whatsoever there is a small amount which can be helpful and these items will be described and discussed in the next chapter.

CHAPTER 8
EQUIPMENT

THE basic equipment for mammal-watching, and the best, is that which your own body provides—keen eyesight, alert hearing and a sensitive nose. These can be improved to some extent by practice and enthusiastic interest is the whetstone which sharpens these natural tools.

A few scientific aids can be helpful. A folding rule or a tape measure is useful, for example, in checking body and tail length of a dead small mammal. As many mammals are nocturnal some illumination at night is virtually essential. The light from an electric torch would scare mammals away but if some transparent red paper is fastened over the glass they do not seem to be bothered by soft red light although of course the brightness is diminished. For finding out what small mammals are in a particular area without destroying them in a break-back trap the Longworth mammal trap is invaluable for catching them alive. A number are really needed to place at a regular distance apart but this of course costs money and the purchase of one will do for a start, others can be added when you can afford them. There are other cages for the live trapping of larger mammals such as rats, squirrels and mink but the small mammal trap is the most useful for a beginner. Animal-attractant ointments can be obtained and this may be useful in masking human scent as well as enticing animals to places where you want them.

Binoculars are a must unless you have exceptionally good eyesight and even then they are still very useful. Again we must remember that many mammals are active in conditions of poor light at dusk so that this must influence our choice of binoculars. The main requirement will therefore be good illumination, and as a bonus, a pair with as wide a view as possible. These are found in the type which are sometimes called night glasses. On them will be marked the sign '7×50'. This means that they magnify seven times and have objectives which are 50 mm in width. The objectives are the lenses at the opposite end from the eye-pieces and their function is to gather in the light. These binoculars have good light-gathering power, a moderate stereoscopic effect

31. Trap for catching rats alive

which makes the animal stand out in relief, and a fairly wide field of view. The magnification is only of medium power but should be adequate for mammal-watching. If it were increased, other factors meanwhile remaining the same, both the field of view and the illumination would be reduced so that a moderate magnification must be accepted.

You will hardly need me to tell you that the photographing of wild mammals is inevitably a difficult matter. It is a tough enough assignment to look at mammals without getting near enough to obtain a photo. It is bound to distract some of your attention from mammal watching and for that reason and because it is difficult to obtain an acceptable picture, I do not recommend this for beginners. Nevertheless it may be that some of you are already interested in photography and eager to make the attempt.

EQUIPMENT

It may be best to begin in a modest way by photographing a set of tracks. A good set must first be found; thin, crisp snow is an ideal medium. To make the trail show up clearly in the picture the best time to take this kind of photo is in the early evening when the sun is low; the shot is taken against the light so the use of a lens hood is essential. For an artistic result, the trail should be aligned diagonally across the photo. If you are making a collection of photos of this type it will be necessary for you to ensure that you stand the same distance away from the footprints on every occasion; this will increase the scientific value for it will mean that the photos can be compared with each other because they are all on the same scale. Other relatively easy subjects are the fraying of trees by deer, evidence of an animal's feeding habits and the holes of burrowing mammals. Almost any simple kind of camera will be suitable.

To photograph a mammal itself, however, the best type of camera is a 35 mm with interchangeable lens because a telephoto lens will usually be required. It is best to begin by photographing animals in a wildlife park where you can gain experience in semi-natural surroundings. After a while you may like to try your hand at mammals in the wild. Do not forget the possibilities that may occur in your own garden. The photo of the rat coming through the hedge to raid the refuse sack was hurriedly taken from my kitchen window with a 135-mm telephoto. This low power was adequate for the purpose because the kitchen served as a hide. Usually, however, a more powerful lens is needed, say 300 or 400 mm. The reason for this is that it is seldom possible to get at all close to a wild mammal. You might almost tread on a sleeping fox but it would not stop to be photographed! A 135 mm would suffice at a badger sett and a standard lens might even be used. I have used a 50-mm lens to photograph a shrew which I had trapped on an Orkney beach. A plastic bowl, quickly decorated with a few stones and some vegetation, held the animal long enough for me to take several acceptable colour transparencies. But these are exceptions. Most mammals will need to be stalked and since the photos will almost invariably have to be taken from a distance a powerful long focus lens is essential even for large creatures like deer.

When stalking you are not likely to have time to fix up a tripod and change lenses. It is therefore advisable to carry the telephoto already in position on the camera. This should be carried carefully with one hand supporting the end of the lens, otherwise the weight may cause damage especially if it is a bayonet attachment. It will help if you can take an exposure meter reading at the outset, setting the aperture and shutter speed accordingly. If there is an obvious change in the lighting conditions another reading will have to be taken but otherwise all that will need to be done when a mammal is sighted is to aim, focus and shoot. Not having a tripod can be a problem but there is a device called a monopod which is helpful for keeping the camera steady. Alternatively on some occasions you may be able to rest the camera on a fence post or support it against a tree trunk.

If you can find a convenient tree perch near a sett entrance you may be able to obtain a photo of a badger. Do not expect to do this first time; a number of attempts may be necessary. Flash, of course, will be essential. Many of the woodland setts with overhanging trees are in a dark situation even in daytime and powerful illumination will be needed. A flash bulb gun can be used with a large bulb. It is possible to obtain a gun which fires two bulbs simultaneously. There are a couple of points to remember. Flash bulbs will not 'freeze' a fast-moving animal and for this electronic flash would be required. Secondly, guide numbers are based on a hypothetical situation of a shot taken in a room with some light reflecting off the wall surface, so that outdoors more light will be needed. This is largely a question of experiment by trial and error. If the situation is an exposed one the animal may not accept you as close as the camera needs to be; in this event it is possible to purchase an extended cable release which will enable you to stay a safe distance away. A more elaborate device is the setting up of a trip wire for nocturnal mammals on a known and well-used path. But I repeat, unless you are already a keen photographer I do not advocate mammal photography for beginners.

What I have just written applies with equal force to sound recording. Nevertheless, because some may already have experimented with recording subjects other than natural history ones, brief mention of the possibilities is made now. The problems are rather different from those of photography and some might say that they are more difficult although the recording enthusiast would probably deny this. One common denominator at least they possess: endless patience is an indispensable requisite for both.

There are first of all problems inherent in the nature of the subject.

32. *The brown rat is an opportunist feeder; here he raids a refuse sack*

Many mammals are silent for most of the time and some of those that are very vocal are too soft and high-pitched to record satisfactorily. Mammal sound recording is inevitably a restricted field but there are some sounds that are well worthwhile attempting to record. Dog fox and vixen answering each other on a winter's night; badger cubs at play; a hedgehog quarrel; seal music and a red stag or fallow buck defending his harem.

Merely to list out the possibilities is enough to create enthusiasm. Unfortunately there are technical difficulties standing in the way. It is one thing to record a voice or instrument indoors; it is quite another to record a mammal in the wild. Most people for general purposes use a mains recorder; outdoors, of course, a battery-operated model is essential. The position is further complicated by the fact that for general use cassette recorders are replacing the reel to reel type and it is a battery-operated, reel to reel recorder which is much to be desired for nature recording. Only a few kinds are manufactured now and they are not cheap. Battery-operated cassette models are cheaper and you may feel that one of these will have to satisfy you. The guidance of a helpful shop manager who is known to possess expertise in this field and on whose integrity you can rely would be invaluable in helping you to select a model which is the best for the purpose having regard to the money you are prepared to spend. It must be clearly understood, however, that unless you can afford the very best you will not get the best quality of performance.

In addition to the actual recorder a microphone is needed and a parabolic reflector is virtually essential in order to concentrate on and isolate the particular sound required. The microphone is inserted in the reflector and then aimed in the direction of the sound. Here again, there are different types of microphones and unless you already possess the relevant technical knowledge skilled advice will be needed. One fact which will influence your choice is whether in the main you wish to compile a library of mammal sounds, as pure as you can get them, or whether your interests are aesthetic and you are happy to include other pleasant associated sounds such as bird song in order to evoke the authentic atmosphere of the habitat. If the former is the case, then you will require a one-directional kind of which the gun microphone is one example.

Of course you will want to exclude undesirable extraneous noises like motor traffic or distant trains and you will need to train your ears to listen for these sounds. The human ear is very selective and often fails to take in that which the microphone faithfully picks up, sometimes with disastrous results for the recording. Much practice will be needed to bring satisfactory results. Wind is the great enemy and a wind shield is an essential piece of equipment. Early morning will be the best time for most subjects and the wind is often less at that time also. Headphones are an optional item although they are very useful in some circumstances such as at night when the sound may not be easily located. Recordists often play a recording of the animal they wish to record again in order to attract it near the microphone and raise an answering voice. This is an experiment well worth trying although it may not always produce the desired effect.

CHAPTER 9

MAMMALS AND MAN

MEN have long held a curiously mixed-up attitude towards the wild mammals of the countryside. Without a doubt primeval man often had occasion to react with fear: on dark nights when he heard the howling of the wolves deep in the dense forests; or when he rounded a bend to see a menacing brown bear a few paces away; when he was confronted by the tusks of a snarling wild boar at bay or when he was pursued by the pounding hooves of a bull aurochs. As a primitive agriculture began to develop, to the feelings of fear were added distrust and anger as animals ranging from mice to deer, raided his hard-won crops. Balanced against these negative emotions have been several more positive ones. An instinctive pleasure in pursuing animals for sport; a reluctant admiration for the courage of the larger carnivores when attacked by hounds, even a sneaking affection for the cunning of Reynard the fox.

This welter of contradictory feelings is still prevalent today. I realize that they are present in myself. The screams of badger and fox heard when alone will always momentarily alarm me. Anger will be dominant when mice eat my sowings of peas and moles uproot seedlings. On the other hand, I cannot but admire the impudence of the squirrel raiding my refuse sack and the courage of a weasel run over by a car rearing up to attack me with its last dying gasp. I like to see the gracefulness of the roe deer and the nobility of the red stag. Whether they exasperate or evoke affection, mammals never cease to fascinate.

Pests

It must be admitted that the viewpoint of country people, especially farmers, is often governed solely by economic considerations. I mean, in relation to wild mammals. This attitude is understandable. Of course economic factors must be taken into account and sometimes control measures must be undertaken. Unfortunately some occupiers of land do not bother to enquire if a particular mammal species is causing damage but too readily assume that all mammals are destructive to their interests so that they are inclined to destroy all mammals on sight. But an animal species in its country of origin has a right to exist; it was not created for man's economic benefit. To eradicate a species is a crime against civilization. In this section we look at mammals which are regarded as pests.

In the small mammals, bank and field voles and field and house mice are undoubted pests to agriculturalists. Bank voles also cause damage in plantations and I have seen Lawson cypress trees killed by the ring-barking of this animal. Ring-barking is the name given to the removal of a strip near the base of a tree; when the ring is complete, the tree dies. Field voles are liable to extreme fluctuations in numbers and vole plagues cause serious damage to grassland.

Rats are a menace not only because of the immense damage they do but also because they are carriers of several dangerous diseases. One of these is called Weil's disease which in man frequently causes death; it has been calculated that nearly 50 per cent of rats carry this organism. The black rat can transmit bubonic plague, otherwise known as the Black Death.

Rats are so widespread and abundant that successful control is very difficult. Local authorities employ pest-control officers and there are private firms which specialize in this work. There are two major problems. One is that if the same type of poison is continually used a resistance is sometimes built up in the rat population. This has happened in a certain area of the Welsh border counties where a much-used poison was found to be not having so much effect. This is a common problem with chemical control and occurs with insecticides as well. The second difficulty is that rats are so numerous that they occupy a habitat to its capacity. If some are destroyed others move in to take their place and this coupled with rapid breeding means that the attempt at control can be a waste of time. It is clear therefore that to be successful rat-control measures must aim for as near a total kill as possible and over as large an area as possible.

It is not easy to present a clear case for or against the mole. They do good and harm. They aerate the soil and consume noxious insects but, on the other hand, they eat large quantities of worms which are beneficial to the soil and they can damage meadows with their runs and mole-hills. In general, they probably do little damage to a farmer's interests unless too many are concentrated in a small area. Mole-catching was a full-time employment for some countrymen up to comparatively recent times. Some men probably still work part-time at this but I doubt whether any are engaged solely in this occupation today. A few years ago I saw a mole-catcher's gibbet, a long line of mole carcasses suspended on a fence at a Scottish farm. Mole clearance is undertaken nowadays by pest control organizations.

Until the onset of myxomatosis the rabbit was probably the most serious agricultural pest apart from the rat. In many areas they are returning and although periodic bouts of the disease recur they can still be a problem in some localities. The reduction of the rabbits' traditional enemies, the stoat, weasel, fox and buzzard by game-preserving interests, helped to increase the population. By the early 1950's it has been estimated that the rabbit population in Britain approached 100 million. A survey made in 1952 estimated that about $1\frac{1}{2}$ cwt of grain per acre could be lost.

The Agriculture Act, 1947 required land occupiers to control rabbits. These powers were extended by the Pests Act 1954 by which particular localities could be designated as rabbit clearance areas. In 1958 rabbit clearance societies began to be formed. Rabbits are destroyed in several different ways; often the most successful way is that which is used by pest officers, the gassing of rabbits in their burrows.

The brown hare is a minor pest of agriculture. In small numbers they are no problem but when they become numerous they need to be controlled. Shooting hares requires more skill than shooting rabbits due to the tremendous speed of these animals. Many mountain hares live well above agricultural ground but they are regarded with disfavour by Scottish farmers because of their habit of descending to lower ground in winter and feeding on arable crops. They probably cause more damage in plantations by their practice of barking trees and eating the tops out of young conifers.

The grey squirrel is a serious pest to farmer and forester. They strip off bark not as the rabbit does, to eat it, but in order to get at the sweet sap underneath. Further damage to trees is caused by the eating of buds. On agricultural land corn is amongst the food consumed in late summer. From the 1930's onwards campaigns were organized against this mammal and in 1937 the Grey Squirrel (Prohibition of Importation and Keeping) Order was promulgated. Squirrels can be caught in traps particularly in the autumn and winter but shooting has been a favourite method. To assist the shooters, dreys are poked with sectional poles which can be fastened together to a maximum of some 18 metres (60 feet) in length. Despite the relentless war which foresters have waged against this mammal it is still as numerous as ever and in some quarters it is now considered pointless to try and exterminate a population in any one wood since the surplus from adjacent woods simply moves in and the last state is the same as the first.

Coypus do both good and harm to river interests. They help to keep waterways clear but on the other hand, they do damage by burrowing in the banks. They also eat sugar beet and their feeding activities spoil the appearance of reed beds. It is generally accepted that like the grey squirrel total elimination is impossible and the aim is to contain them as far as possible in the existing breeding area. If you are in East Anglia you may see vans which are marked Coypu Control and these belong to an organization set up solely for the purpose of trapping these mammals.

Trapping is by the same means used to keep down mink which are pests for a different reason, being voracious carnivores. One night a Hampshire garden abutting on a stream was visited by a mink which raided the poultry house and killed seven ducks, six chickens and one goose only to leave them there. Occupiers of land in England and Wales who know that they have mink on their land are required to notify the Ministry of Agriculture and to take all possible steps to eradicate the animals.

Foxes are not popular with farmers because of their propensity for taking poultry and the occasional lamb. A relentless war is waged against them and an estimated 50,000 are killed annually. Despite this toll, the species thrives everywhere. Coastal foxes are a menace to terneries where they pay nocturnal visits to eat eggs and young. A more serious danger from the fox has arisen in recent years with the realization that the fatal disease of rabies, which is endemic on the Continent, could easily be transported to this country. In Europe it is spreading westwards towards the Channel ports. Rabies is caused by a virus which infects the brain and is transmitted to humans by a bite from the infected animal. Unfortunately in its maddened state it loses its natural fear of man and rushes to bite the nearest human beings. Foxes are known to be important carriers of the disease and the high population

33. *Badger gate erected by the Forestry Commission to allow the animals easy exit and entrance to the plantation*

in Britain's dockland gives cause for concern.

The badger has been a generally harmless creature in the countryside and there has been no justification for the persecution which has sometimes occurred. It was most unfortunate that at about the time legislation was enacted to protect the badger, some animals were found to be carriers of tuberculosis, chiefly in parts of the West County. Although the Badgers Act, 1973 gave them a measure of protection the Conservation of Wild Creatures and Wild Plants Act, 1975 gives Ministry of Agriculture officials power to gas badger setts where animals are suspected of harbouring the disease. On one 3000 acre Dorset farm over 600 cwt of gassing powder is reported to have been used to destroy badgers. That some badgers in some localities have TB is a fact which cannot be disputed. So far as I know, however, it has not yet been conclusively proved that the badgers, at least in some instances, were not themselves victims with the cattle, having got the disease from another source. An occasional rogue badger will kill chicken but this is comparatively rare and when they do, unlike the fox, they do not kill more than they eat.

The increase in the deer population since the Second World War has brought a corresponding increase in damage to field crops and forestry plantations. Some of this damage has already been referred to in the chapter on food. Deer cannot be eliminated even if that were desirable but various steps can be taken to minimize the damage. Selective shooting of animals, that is, those causing the most damage and those which are poor specimens; planting where possible trees which are not so acceptable to deer; leaving shrubs for bucks to fray rather than commercial trees; these are some of the ways in which deer can be reconciled with forestry.

There remain the seals. Probably partly as a result of the close season protection afforded them, seals have increased in numbers. The Government has now authorized the killing under licence of a certain number of pups during the close seasons and culls have been made in the Wash, the Farne Islands and the Orkneys. This has aroused keen controversy and strong emotions amongst animal lovers, to whom the seal pups are attractive creatures, and the debate continues.

Sport

Countrymen, when they can, turn necessity into enjoyment and make a sporting event out of killing some of these mammals. In the days before myxomatosis a small shooting party with a bag of ferrets would have a good day's sport with rabbits. This animal was a favourite target of the many poachers who followed their illegal pursuits, some for the meat and some just for the kicks they got from outwitting the gamekeepers.

Hares, too, are shot in organized drives but the practice is not regarded by shooting men as being much of a sport since it is not always easy to kill the animal outright. More sport is obtained by hunting. Hares are hunted from October to March chiefly by packs of beagles with hunt staff on foot. Coursing is a quite different process. It consists of the running of a pair of greyhounds who hunt by sight and not scent. Points are awarded to the dogs by a judge and the best performing dog is not necessarily the one which kills the hare.

The majority of hunts in Britain have the fox as their quarry. Foxhunting is a long-established country tradition which goes back at least to the early 18th century. It has been the subject of keen controversy in recent years on ethical grounds; there have been debates in Parliament and a bill to make fox-hunting illegal was introduced but later withdrawn. More will undoubtedly be heard on this matter in the future. In the lowlands fox hunts are mounted but in hilly country as in the Lakeland fells hunting is carried out on foot. Otters have been hunted by a small number of hunts, about a dozen in Britain, but the scarcity of the quarry forced them to relinquish this pursuit and it is now in any case illegal. Fallow deer are hunted in the New Forest but in a few places such as Exmoor and Norfolk the quarry are red stags.

In the Scottish Highlands deer-stalking with rifles is no longer the sport just of a few landowners and their guests but is becoming a tourist industry attracting visitors from overseas. After the Second World War poaching increased considerably, particularly in Scotland, although it occurs in other places such as the New Forest. Much cruel suffering has been caused to deer by the use of shot-guns and crossbows which have not killed the animals outright.

Another barbarous practice which has been indulged in the West Country is badger-baiting. This consists of digging out a badger from a sett and then putting dogs on to attack the inoffensive creature. Mild-natured it is until aroused but then it is able to give a good account of itself until it is cruelly despatched. It is hoped that conservation measures and the climate of public opinion will eliminate this so-called 'sport'.

Commercial Value

To primitive man wild mammals were an important source of food and clothing and they were pursued with stone weapons wielded by Palaeolithic hunters. From Neolithic times onwards with the domestication of certain mammals the killing of wild stock became less important. Nevertheless, even in Britain today they have some economic use and as we shall see in a moment, the process of domestication is not ended.

Hedgehogs have no economic value but they have for long been a favourite food of gypsies. No-one has a good word for rats; no-one, that is, except scientists for whom the rat is a valuable laboratory animal for such purposes as the testing of medicines. Rabbits were once an important source of food and in the last century on the downlands of southern England they were, in effect, farmed in vast warrens. With the onslaught of myxomatosis rabbits as food became distasteful to the public. Hares are good to eat but are not highly valued in Britain. Moleskins have been put to good use for many centuries; waistcoats, caps, bed-covers and purses are some of the items produced. There was still a commercial interest in moles and skins were exported in the first few years after the Second World War but nowadays in Britain moleskins are no longer a commercial proposition.

Otters used to be trapped for their fur especially in northern Britain and one or two of the old traps still remain in existence in the Shetlands. In these northern isles seals are occasionally hunted for their skins which are used for clothing and various items of an ornamental nature. As well as contributing to the tourist trade, deer supply skins for rugs, wallets and other objects; various ornamental items such as table lamp stands, knife handles and the like are also made from antlers. The latest development is the establishment of red deer farms in one or two places but it is early days yet to say whether this idea is going to spread.

Conservation

It is only comparatively recently that another way of looking at wild mammals has come into prominence. This approach regards them both as interesting animals to study and as worthy of conservation as an integral part of Nature. Some species must be controlled but they do not have to be exterminated. There is no doubt that this viewpoint not only deserves its place as a justifiable way of looking at mammals but that it is a better way than regarding mammals solely as pests, as objects for sport or for economic use.

In the early years of the present century it was thought that the Atlantic grey seal was on the way to extinction. Certainly the Hebrideans used to hunt these mammals at their breeding stations. Parliament was persuaded of the need for preservation and passed the Grey Seal Protection Act, 1914, which provided a close season during the autumnal breeding period from 1st October to 15th November; this was enlarged in 1932 to cover the period from 1st September to 31st December. Since then there has been a considerable increase in their numbers although it is possible that there are other contributory factors apart from protection. Despite its relative abundance its habit of breeding on a few islands makes it especially vulnerable to those who see it either as a threat to

34. *An old otter trap in the Shetlands*

their livelihood or as a means of making money. It ought not to be forgotten that the waters round the British Isles hold the majority of the world's population of these mammals and that therefore we have a special responsibility to conserve them. Much interest developed in this subject during the 1960's and it was realized that another Act of Parliament was required. In 1970 the Conservation of Seals Act was passed which covered common as well as grey seals; protection was given to both species in their differing breeding seasons but allowance was made for the issue of licences for controlled culling of seals outside of the National Nature Reserves.

Laws for the protection of deer were made by the Norman Kings but this was solely in the interests of their sport and not for conservation as we understand the term. It was not until 45 years after the first Grey Seal Act that another species of mammal was protected by law. This, the Red Deer (Scotland) Act, 1959 only applied in all its aspects to one species of deer in one part of Britain but it was at least a start. It originated partly because of the necessity to take action against poachers. The provisions of the Act against poaching applied to all species of deer in Scotland. But the Act is primarily concerned with red deer and under it the Red Deer Commission has been established to give an overall control to the large herds of deer which roam over many square miles of the Highlands. The aim is control not extermination, and separate close seasons have been provided for stags and hinds. Stags can only legally be shot in the late summer and autumn and the hinds only during the winter months. The Act also made provision for close seasons for other deer to be brought in at a later date and regulations were made giving protection to fallow, roe and sika in 1966. Three years earlier the Deer (England and Wales) Act, 1963 came into force providing close seasons for the foregoing four species in England and Wales.

After some abortive attempts at legislation to protect badgers the Badgers Act became operative on 1st January, 1974. It arose from a desire to protect this fine mammal from acts of cruelty. Landowners, however, are still entitled both to kill badgers and to authorize other people to do so. On the other hand, it is now illegal to sell a badger or to keep one as a pet. The future of the badger is now clouded with uncertainty since the discovery that some are carriers of TB. It is to be hoped that these outbreaks can be confined to localized areas.

A notable step forward was taken with the Conservation of Wild Creatures and Wild Plants Act, 1975. This Act provided for the protection of a number of rare plants and animals. Initially just two mammals were included, the greater horseshoe and mouse-eared bats but since 1st January, 1978, the otter has been added to the list. It is illegal to kill, capture or mark them in any way; the Act also prohibits the ringing of any species of bat except under licence from the Nature Conservancy Council.

Apart from Acts of Parliament the establishment of National and Local Nature Reserves gives localized protection to the mammals which live therein. One vital issue remains. Admirable though the various protection Acts may be, how effectively can they be enforced? Policemen are not required to be naturalists nor are they given to wandering over fields and through woods on their beats. Many offences are likely to be committed unseen on private land to which the public have no access. Clearly enforcement is a weak link in the chain of mammal protection. The ideal solution is for more and more people, especially landowners, to become interested in mammals for their own sake. A climate of opinion will then be established whereby mammals are treated with every consideration and, when necessary, humanely destroyed.

INDEX

Aurochs, 21

Badger, 10, 11, 16, 41, 52, 53, 73, 74, 75, 76, 84, 85, 88, 90, 93, 98, 99, 101, 105, 119, 121, 124

Bats, 52, 102, 104
 barbastelle, 36
 Bechstein's, 35
 Brandt's, 34
 Daubenton's 35, 57
 greater horseshoe, 32, 34, 68, 124
 grey long-eared, 36
 Leisler's, 35, 36
 lesser horseshoe, 34
 long-eared, 36, 55, 68
 mouse-eared, 32, 35, 124
 Natterer's, 35, 58, 68
 noctule, 36, 68
 pipistrelle, 36, 55, 87
 serotine, 35, 68
 whiskered, 34, 68

Bear, brown, 21
Beaver, 23
Binoculars, 109
Bison, 21
Boar, wild, 21, 22

Cat, Scottish wild, 18, 42, 56, 71, 78, 80, 85, 101,
Coypu, 24, 40, 58, 66, 67, 78, 84, 99, 118

Deer,
 Chinese water, 27, 45, 50, 84
 fallow, 26, 44, 50, 64, 83, 86, 89, 91, 96, 100, 121, 124

muntjac, 27, 45, 50, 64, 84, 86, 89, 96
 red, 18, 43, 48, 55, 64, 82, 89, 91, 96, 100, 107, 121, 122, 124
 roe, 10, 44, 50, 64, 83, 84, 86, 91, 96, 101, 108, 124
 Siberian roe, 27
 sika, 26, 43, 50, 64, 83, 89, 96, 101, 108, 124
Dormouse,
 common, 40, 53, 66, 80, 102
 edible, 26, 40, 53, 55, 84

Echo-location, 88
Elk, 22

Fox,
 arctic, 14
 European, 16, 40, 47, 74, 76, 80, 84, 85, 88, 90, 96, 97, 100, 101, 105, 118, 121

Goat, feral, 27, 45, 63, 64

Hare,
 brown, 17, 36, 37, 54, 66, 78, 80, 81, 84, 88, 90, 98, 102, 104, 117, 120, 122
 mountain, 14, 18, 37, 55, 66, 78, 86, 88, 91, 98, 99, 102, 117, 120
Hedgehog, 16, 30, 46, 67, 68, 81, 87, 90, 93, 99, 103, 121

Mammoth, 14
Marten, pine, 40, 56, 69, 78, 106
Mink, 24, 41, 59, 71, 78, 88, 118
Mole, 16, 30, 54, 67, 77, 82, 84, 103, 116, 122

INDEX

Mouse,
  harvest, 38, 54, 66, 78
  house, 38, 54, 55, 116
  long-tailed field, 10, 20, 38, 54, 55, 66, 78, 102, 103, 116
  St. Kilda house, 23
  yellow-necked, 17
Musk-ox, 14

Otter, 16, 41, 59, 60, 71, 78, 85, 88, 90, 93, 102, 121, 122, 124

Photography, 110–12
Polecat, 41, 56, 71, 78, 81, 88, 99, 100

Rabbit, 16, 24, 36, 47, 66, 70, 76, 80, 86, 88, 90, 98, 102, 117, 120, 121
Rat,
  black, 24, 38, 55, 75, 87, 99, 116, 121
  brown, 24, 47, 58, 75, 82, 99, 116
Reindeer, 14, 22, 23, 45
Rhinoceros, woolly, 14

Seal,
  Atlantic grey, 42, 60, 62, 72, 80, 81, 84, 87, 89, 90, 120, 122, 123
  common, 42, 60, 62, 72, 80, 81, 84, 87, 89, 90, 120, 122, 123

Sheep, Soay, 27, 45, 63, 64
Shrew,
  common, 18, 31, 54, 67, 78, 81, 87, 90
  lesser white-toothed, 32, 67, 85, 90
  pygmy, 31, 54, 67, 78, 90
  water, 17, 32, 58, 67, 78, 90
Sound recording, 112–14
Squirrel,
  grey, 10, 25, 37, 51, 52, 74, 75, 80, 82, 84, 85, 87, 88, 90, 99, 102, 103, 117
  red, 17, 37, 51, 52, 69, 74, 75, 80, 82, 84, 85, 88, 90, 99, 103
Stoat, 21, 41, 55, 70, 84, 85, 90, 99, 107

Tracks, 93–9
Traps, 109

Vole,
  bank, 20, 37, 38, 54, 58, 66, 78, 102, 116
  Orkney, 38
  short-tailed field, 20, 29, 38, 54, 66, 78, 116
  water, 18, 38, 58, 66, 78, 81, 104

Wallaby, red-necked, 25, 30, 64
Weasel, 16, 41, 55, 70, 90, 99, 107
Wolf, 23, 40